T0135785

Identification and Fault Diagnosis of Industrial Closed-Loop Discrete Event Systems

Identifikation und Fehlerdiagnose industrieller ereignisdiskreter
Closed-Loop Systeme

Identification et diagnostic des systèmes à événements discrets
industriels en boucle fermée

vom Fachbereich Elektrotechnik und Informationstechnik
der Technischen Universität Kaiserslautern
zur Verleihung des akademischen Grades
Doktor der Ingenieurwissenschaften (Dr.-Ing.)
genehmigte Dissertation

von
Matthias Roth
geb. in Kaiserslautern

D 386

Eingereicht am: 30.07.2010
Tag der mündlichen Prüfung: 08.10.2010
Dekan des Fachbereichs: Prof. Dipl.-Ing. Dr. Gerhard Fohler

Promotionskommission:
Vorsitzender:
Prof. Dr.-Ing. Wolfgang Kunz
(Technische Universität Kaiserslautern)

Berichterstattende:
Prof. Dr.-Ing. habil. Lothar Litz
(Technische Universität Kaiserslautern)
Prof. Dr. habil. Jean-Jacques Lesage
(Ecole Normale Supérieure de Cachan, Frankreich)
Prof. Dr.-Ing. Birgit Vogel-Heuser
(Technische Universität München)
Prof. Dr. Janan Zaytoon
(Université de Reims, Champagne-Ardenne, Frankreich)

Bibliografische Information der Deutschen Nationalbibliothek

Die Deutsche Nationalbibliothek verzeichnet diese Publikation in der
Deutschen Nationalbibliografie; detaillierte bibliografische Daten sind
im Internet über http://dnb.d-nb.de abrufbar.

ISBN 978-3-8325-2709-9

Logos Verlag Berlin GmbH
Comeniushof, Gubener Str. 47,
10243 Berlin
Tel.: +49 (0)30 42 85 10 90
Fax: +49 (0)30 42 85 10 92
INTERNET: http://www.logos-verlag.de

Abstract – The competitiveness of manufacturing companies strongly depends on the productivity of machines and production processes. To guarantee a high level of productivity, downtimes occurring due to faults have to be kept as short as possible. This necessitates efficient fault detection and isolation (FDI) methods. In this work, a model-based FDI method for the widely used class of industrial closed-loop Discrete Event Systems is proposed. The considered systems consist of the closed-loop of plant and controller. Based on the comparison of observed and modeled system behavior, it is possible to detect and to isolate faults. Unlike most known methods for FDI in Discrete Event Systems, this work proposes working with a model of the fault-free behavior rather than working fault models. Inspired by the concept of residuals known from FDI in continuous systems, a new approach for fault isolation based on fault-free Discrete Event System models is developed. The key of any model-based diagnosis method is to have an accurate model of the considered system. Since manual model-building can be very difficult for large industrial systems, an identification approach for this class of systems is introduced. Based on an already existing monolithic identification algorithm, a distributed adaptation is developed which allows treating large, concurrent systems. The key of the proposed approach is an automatic decomposition of a given closed-loop Discrete Event System using an optimization approach which analyzes observed system behavior. The methods developed in this thesis are applied to a mid-sized laboratory system and to an industrial winder to show their scalability.

Zusammenfassung – Die Wettbewerbsfähigkeit von Industrieunternehmen hängt maßgeblich von der Produktivität der eingesetzten Anlagen und Produktionsprozesse ab. Um ein hohes Maß an Produktivität zu garantieren, müssen durch Fehler verursachte Standzeiten so kurz wir möglich gehalten werden. Dazu werden effiziente Methoden zur Fehlererkennung und Fehlerisolierung (FDI: fault detection and isolation) benötigt. In der vorliegenden Arbeit wurde ein modellbasiertes FDI-Verfahren für die weit verbreitete Klasse ereignisdiskreter Closed-Loop Systeme entwickelt. Die betrachteten Systeme bestehen aus dem geschlossenen Kreis von Steuerung und Prozess. Durch den systematischen Vergleich von aktuell beobachtetem und durch das Systemmodell erwartetem Verhalten können Fehler in Echtzeit erkannt und isoliert werden. Im Unterschied zu den meisten aus der Literatur bekannten FDI-Verfahren für Ereignisdiskrete Systeme wird in dieser Arbeit ein Modell des fehlerfreien Systemverhaltens an Stelle von Modellen des Fehlerverhaltens verwendet. Angelehnt an das von kontinuierlichen Systemen bekannte Fehlerisolierungsprinzip der Residuen wurde ein neuer Ansatz zur Fehlerisolierung in Ereignisdiskreten Systemen entwickelt. Der wichtigste Bestandteil von modellbasierten Diagnoseverfahren ist ein genaues Systemmodell. Da die Modellbildung „von Hand" für Systeme im industriellen Maßstab meist zu aufwendig und teuer ist, wurde ein Identifikationsverfahren für die betrachtete Systemklasse entwickelt. Ausgehend von einem bereits existierenden monolithischen Identifikationsalgorithmus wurde ein verteiltes Identifikationsverfahren eingeführt, das für komplexe industrielle Anlagen mit ausgeprägten Nebenläufigkeiten geeignet ist. Kern des Verfahrens ist die automatische Dekomposition eines Closed-Loop Systems in nebenläufige Teilsysteme. Dies geschieht mit Hilfe eines Optimierungsansatzes, der beobachtetes Systemverhalten analysiert und durch Minimierung einer Gütefunktion eine optimale Systemunterteilung erreicht. Die in dieser Arbeit entwickelten Methoden wurden sowohl im Labormaßstab, als auch im Industriemaßstab erfolgreich getestet.

Résumé – La compétitivité des entreprises manufacturières dépend fortement de la productivité des machines et des moyens de production. Pour garantir un haut niveau de productivité il est indispensable de minimiser les temps d'arrêt dus aux fautes ou dysfonctionnements. Cela nécessite des méthodes efficaces pour détecter et isoler les fautes apparues dans un système (FDI). Dans cette thèse, une méthode FDI à base de modèles est proposée. La méthode est conçue pour la classe des systèmes à événements discrets industriels composés d'une boucle fermée du contrôleur et du processus. En comparant les comportements observés et attendus par le modèle, il est possible de détecter et d'isoler des fautes. A la différence de la plupart des approches FDI des systèmes à événements discrets, une méthode basée sur des modèles du comportement normal au lieu de modèles des comportements fautifs est proposée. Inspiré par le concept des résidus bien connu pour le diagnostic des systèmes continus, une nouvelle approche pour l'isolation des fautes dans les systèmes à événements discrets a été développée. La clé pour l'application des méthodes FDI basées sur des modèles est d'avoir un modèle juste du système considéré. Comme une modélisation manuelle peut être très laborieuse et coûteuse pour des systèmes à l'échelle industrielle, une approche d'identification pour les systèmes à événements discrets en boucle fermée est développée. Basée sur un algorithme connu pour l'identification des modèles monolithiques, une adaptation distribuée est proposée. Elle permet de traiter de grands systèmes comportant un haut degré de parallélisme. La base de cette approche est une décomposition du système en sous systèmes. Cette décomposition est automatisée en utilisant un algorithme d'optimisation analysant le comportement observé du système. Les méthodes conçues dans cette thèse ont été mises en œuvre sur une étude de cas et sur une application d'échelle industrielle.

Acknowledgements

The work that resulted in this thesis could not have been accomplished without several persons' assistance, support, and encouragement.

First of all, I would like to thank my doctoral advisors, Professor LOTHAR LITZ and Professor JEAN-JACQUES LESAGE for the opportunity to perform my work at their institutes. I am grateful for their guidance, advice, encouragement and for the intellectual freedom they have given to me during this work. They gave me the opportunity to experience both, the French and the German cultures of research.

Besides to Professor LOTHAR LITZ and Professor JEAN-JACQUES LESAGE, my sincere thanks go to the other members of the evaluation board: Professor BIRGIT VOGEL-HEUSER from Technische Universität München, Professor JANAN ZAYTOON from Université de Reims Champagne-Ardenne, and to the chairman Professor WOLFGANG KUNZ.

During my work, I was four years part of the Institute of Automatic Control in Kaiserslautern. Working at this institute means being part of a team, somehow part of a family. I am grateful to my colleagues for many interesting discussions on my work, on their work, or on anything else which caught our interest! Thank you, dear colleagues, for your friendship, for your honesty and your humor. I also thank my colleagues of the LURPA in Cachan for their kindness and helpful advice. They always made me feel welcome in Cachan. Apart from discussions on my work, they gave me some deep insights into the French way of living and thinking which really broadened my horizon.

Some (former) students also contributed to my work. I thank NELIA SCHNEIDER an MICKAEL DANANCHER for detailed tests of the residual method, THORSTEN RODNER for his contribution to the partitioning approach and MARCO PÖRSCH for his contribution to apply the diagnosis method to an industrial system.

This work was part of an industrial project of the Institute of Automatic Control with Freudenberg Vliesstoffe KG based in Kaiserslautern. I would like to thank all the people who helped to conduct the experimental part of my work. I am especially grateful to Dipl.-Ing. PETER SCHMITT and his successor Dipl.-Ing. ALEXANDER BARNSTEINER for their interest in my work as well as JÜRGEN DERUS for his continuous help and advice.

To MARTIN FLOECK, THOMAS GABRIEL, STÉPHANE KLEIN, KONSTANTIN MACHLEIDT, THORSTEN RODNER, and THOMAS STEFFEN: Many thanks for proofreading the manuscript and hinting at a considerable amount of flaws I have eliminated with your help.

Finally, I give the heartiest gratitude to my family, in particular to my wife ANNA, for their love, their patience and their encouragement.

Kaiserslautern, October 2010

MATTHIAS ROTH

Contents

1 Introduction

1.1 Motivation

An increasing competition among manufacturing companies leads to growing demands on productivity of technical systems such as industrial production facilities. An important issue to increase productivity is to improve the availability of a system by reducing downtimes due to faults. Two main strategies are possible to achieve minimal downtimes. The first one is to perform preventive maintenance with the aim of changing critical system components before a malfunctioning due to wear and tear appears. The limits of this strategy are faults that appear suddenly and are not predictable based on experience with the concerning hardware. The second maintenance strategy is to perform necessary repair actions after a failure[1] has occurred as quickly and as precisely as possible to reestablish a running system. For such concerted repair actions it is crucial to gather precise information about faults in the system by using fault diagnosis techniques.

Various fault diagnosis techniques for technical systems have been proposed by the scientific community. Examples for these approaches are data-driven, knowledge-based and model-based diagnosis methods. In this work, the class of model-based approaches is focused in particular. A method yielding a model-based diagnosis technique for a widely used class of closed-loop Discrete Event Systems (DES) is proposed. The considered systems consist of a closed-loop of controller and plant as depicted in figure 1.1. The plant represents the physical facilities of the system like conveyors, cylinders or valves as well as products treated in the system. In the controller an algorithm is executed implementing a control strategy. The controller uses sensor information given by its input signals to determine appropriate output signals to control the actuators of the plant. In the considered class of systems, the signals exchanged between plant and controller are discrete with only two possible values (binary coding). During a running system evolution like a production cycle it is possible to get information of the system state by analyzing these signals. Industrial closed-loop DES typically have many input and output signals which lead to a significant complexity when analyzing the system behavior.

In the case of model-based diagnosis, the information gathered by capturing the exchanged signals is used to compare the current system evolution with a modeled behavior. Based on the comparison of expected (modeled) and observed behavior it is possible to detect faults and to determine the necessary information to start concerted repair actions. Model-based diagnosis approaches can be divided in two classes. The first class uses models containing the fault-free behavior as well as the behavior in

---

[1]A definition to distinguish between the notions fault and failure will be given in section 2.3

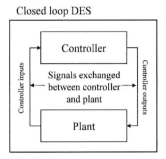

Figure 1.1: Industrial closed-loop Discrete Event System

the case of given faults. If a faulty behavior is observed, it is checked which fault has to be assumed in the model such that it can reproduce the observation. The second class works with models only representing the fault-free system behavior. Using this class, fault detection is based on the assumption that the fault-free model cannot reproduce an observed faulty behavior. If the model cannot reproduce the observation, a fault is detected. The advantage of the second class of approaches is that it is not limited to faults that have been anticipated in the underlying model. The diagnosis approach presented in this work belongs to this second class of model-based methods. The approach addresses the two main aspects of implementing a model-based diagnosis method which are model building and the development of appropriate fault detection and isolation algorithms.

Before a model-based diagnosis method can be implemented, the model has to be built. For industrial systems, model building can be very laborious and expensive due to parameters like system size and complexity. Another general problem when building a system model is that engineers familiar with a given technical system often do not have the necessary knowledge of appropriate modeling formalisms to build models for fault diagnosis. Hence, it is advantageous to automate the model building process as far as possible in order to facilitate the implementation of model-based diagnosis methods. A generic way to obtain models is to use identification methods working on a given data base of observed system evolutions. In this work, identification algorithms are developed leading to appropriate models for online fault diagnosis. The algorithms contain well-defined degrees of freedom which allow adjusting the identified models for diagnosis purposes.

1.2 Contributions of the thesis

The thesis contains two main contributions: Identification of fault-free models for closed-loop DES and fault detection and isolation procedures for diagnosis using the identified models.

Identification: In (Klein, 2005) an identification method for closed-loop DES is presented. It delivers a monolithic automaton modeling the fault-free system behavior. The identification algorithm works with a free parameter k to balance model size and model accuracy. The first contribution of the present thesis is a reformulation of the algorithm of (Klein, 2005) to be less restrictive. Guidelines are developed to properly choose the free identification tuning parameter to get an appropriate model for fault diagnosis purposes. It is shown that the monolithic approach comes to its limits when systems with a high degree of concurrency have to be treated like it is often the case in industrial systems. For these system, using the identified monolithic automaton for fault diagnosis leads to an unacceptably high number of false alerts. To overcome this problem, a distributed identification approach is presented that consists of building partial models for different subsystems. It is explained how the resulting automata network significantly reduces the number of false alerts and how the partial automata can be synchronized to improve their fault detection capability. Since determining appropriate subsystems is a demanding task, an approach to *automatically* find an optimal partitioning is developed. The partitioning approach uses meta-heuristic optimization methods to minimize an objective function representing desired characteristics of the system to be partitioned.

Fault detection and isolation: A literature review revealed that most of the existing model-based diagnosis approaches for DES are based on models containing fault-free *and* faulty system behavior. To use the identified fault-free models for fault diagnosis of closed-loop DES, a fault isolation strategy has been developed inspired by *residuals* known from diagnosis in continuous systems. Generic fault symptoms for the considered system class are presented and in a first step formalized for the monolithic model. Based on these formalized fault symptoms, it is possible to isolate faults in the considered class of closed-loop DES. In the case of fault diagnosis with the identified automata network, it is shown that the distributed models not only reduce the number of false alerts but also have a poorer fault detection capability than the monolithic automaton. Hence, a behavioral tolerance is integrated into the identified automata network as an additional degree of freedom to systematically balance fault detection capability and number of false alerts. The generic fault symptoms are then formalized for the distributed models and the additional degree of freedom which allows fault detection and isolation in the distributed framework.

1.3 Organization

In chapter 2 a literature review on the state of the art in diagnosis and identification of DES is conducted. After a short introduction to DES theory different diagnosis techniques are analyzed. Special emphasis is given to model-based diagnosis methods for DES. Two classes of approaches are presented: Methods working with models containing the fault-free system behavior as well as the behavior in case of given faults on the one hand, and approaches working with fault-free models only on the other hand. The literature review also focuses on the question how to obtain appropriate DES models for fault diagnosis. Recent developments in the field of identification of DES are evaluated.

Since the thesis is based on some important results from (Klein, 2005), Klein's monolithic identification approach is presented in chapter 3. The behavior of the considered closed-loop DES is defined using a formal language. This language is the data base for the identification of a monolithic automaton. Some improvements of the identification algorithm are proposed and several guidelines for determining reasonable values for the free identification parameter are developed. In order to show the limits of the approach, a case study is treated. The considered laboratory facility is introduced and serves as reference application in the following chapters. Based on the case study it is shown that in some cases the monolithic model leads to an unacceptably high number of false alerts.

In chapter 4 it is shown that the number of false alerts can be significantly reduced by dividing a system into concurrent subsystems. It is explained how distributed identification of partial automata building an automata network can be carried out. A formal proof shows that under some assumptions the behavior of the identified automata contains the complete fault-free system behavior although it had not (yet) been completely observed. This leads to a reduction of the number of false alerts. Implications of the results for real systems are shown by means of the case study introduced in chapter 3.

For the identification of partial automata in chapter 4 it is necessary to determine appropriate subsystems. In chapter 5 it is first shown how such a system partitioning can be achieved manually. Since the necessary system knowledge for this approach is not always available, a method to automatically perform an optimal partitioning based on meta-heuristic optimization algorithms is developed. It is shown how desired characteristics of the partitioned system can be approximated giving formal objective functions for the optimization algorithms. An approach to systematically integrate limited physical system knowledge in the automatic system partitioning is also presented which allows improving the quality of the resulting partition. Results of the approach are given for the case study introduced in chapter 3.

Chapter 6 presents fault detection and isolation procedures for the monolithic and for the distributed models. In the first part (section 6.1), the monolithic model is considered. Probabilistic measures to assess the fault detection capability of a given model are introduced. Generic fault symptoms are developed and formalized inspired by residuals known from diagnosis in continuous systems. Using these residuals it is possible to isolate faults having occurred in the system. Several examples for fault detection and isolation are treated for the case study from chapter 3. In section 6.2 the results from chapter 6.1 are systematically adapted for the identified distributed models. For several faults introduced in the case study system the performance of the method is shown.

In chapter 7 the presented methods are applied to an industrial production facility. The results show that the combination of identification and diagnosis methods allows implementing an efficient fault diagnosis system even for large systems with reasonable efforts.

The thesis is concluded by a summary and an outlook in English and extended summaries in French and German.

2 Existing diagnosis and identification approaches for Discrete Event Systems

2.1 Industrial closed-loop Discrete Event Systems

In this work the fault diagnosis and model identification problem in a widely used class of industrial closed-loop systems is addressed. The considered systems consist of controller and plant. In the plant, a set of discrete sensors measures certain process values and delivers them to the controller using the controller inputs. The controller executes a control algorithm and determines appropriate discrete actuator settings in the plant. Commands to actuators in the plant are transfered via controller output signals. Typical sensors and actuators in an industrial closed-loop DES are discrete position or level sensors as well as pneumatic or hydraulic cylinders and conveyor belts.

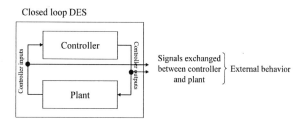

Figure 2.1: Industrial closed-loop Discrete Event System

The behavior of such systems can be observed by analyzing the signals exchanged between controller and plant. In the considered class of systems these signals are binary. From an external point of view, a change in value of a signal can be considered as an event. The occurrence and the order of these events are determined by the system dynamics.

For the implementation of efficient fault diagnosis methods for the considered system class it is necessary to have a formal description of the system behavior. In the next section it will be shown how a sequence of events (changes in value of signals) can be interpreted as a formal language. It will be explained how the dynamics of a given system can be represented using two established formalisms: Automata and Petri nets. After this introduction to DES, several existing diagnosis and identification approaches are presented and evaluated.

2.2 Introduction to Discrete Event Systems

2.2.1 Describing a system by its language

Discrete Event Systems (DES) are a widely used formalism to describe the behavior of systems from different technical domains like manufacturing, transportation or communication. According to (Cassandras and Lafortune, 2006) a DES is defined as follows:

Definition 1 (Discrete Event System, (Cassandras and Lafortune, 2006)). A *Discrete Event System* is a *discrete-state, event-driven* system, that is, its state evolution depends entirely on the occurrence of asynchronous discrete events over time.

An example of the evolution of a DES is given in figure 2.2. It can be seen that the system is in different states at different times and exhibits an event trace. The events occur asynchronously and can (but do not have to) lead to a new DES state. The set of event traces that can be exhibited by a DES is called the system language.

Definition 2 (Language, (Cassandras and Lafortune, 2006)). A *language* defined over an event set E is a set of finite-length strings formed from events in E.

Figure 2.2: Evolution of a DES

According to (Cassandras and Lafortune, 2006) the language generated by a DES can be considered with different levels of abstraction. The most detailed level is to describe a system with a *stochastic timed language*. Determining this language for a given DES necessitates deep system knowledge like probability distribution functions to express the time between successive event occurrences. The stochastic timed language contains all possible event paths of a system with statistical information about them. A system description with a higher level of abstraction is given by the *timed language*. Compared to the *stochastic timed language* it does not contain any statistical information about the event traces. The highest level of abstraction is to use the *untimed language*. It consists of all possible event traces of a system without any timing information and thus mainly delivers information about the order of event occurrences. This work is focused on *untimed language* to describe a DES behavior. Hence the notion *language* always refers to the *untimed language* unless other meanings are explicitly indicated.

In technical domains the language of a given system is usually not known and must be approximated using efficient formalisms. In the next two sections two of these formalisms are introduced: Automata and Petri Nets. Both are capable of generating a language and thus can be used to model a given system. These system models can be constructed manually by considering causal relations of physical components or by an analysis of observed sample event traces. The second approach is referred to as identification.

2.2.2 Automata

Automata are an efficient formalism to describe the language of a DES. Unlike in system theory for continuous systems, in the DES domain there does not exist a standard system description like the state space model using differential equations (Franklin et al., 2001). According to its precise use, many authors propose adapted models for different purposes (e.g. (Alur, 1999) for verification with *timed* automata or (Schroder, 2002) for fault diagnosis with *stochastic* automata). Nevertheless, the concept of automata can be shown using the following basic definition often referred to in literature:

Definition 3 (Deterministic automaton, (Cassandras and Lafortune, 2006)). A Deterministic Automaton, denoted by G, is a six-tuple

$$G = (X, E, f, \Gamma, x_0, X_m)$$

where X is set of states, E is the finite set of events associated with G, $f : X \times E \to X$ is the transition function: $f(x, e) = y$ means that there is a transition labeled by event e from state x to state y; in general f is a partial function on its domain. $\Gamma : X \to 2^E$ is the active event function; $\Gamma(x)$ is the set of all events e for which $f(x, e)$ is defined and it is called the active event set of G at x. x_0 is the initial state and $X_m \subseteq X$ is the set of marked (or final) states.

The language generated by the Deterministic Automaton is built by event traces that occur when the automaton performs a state trajectory starting in its initial state x_0. If the state trajectory ends in one of the marked states $x_m \in X_m$, the trace is part of the marked language:

Definition 4 (Languages generated and marked, (Cassandras and Lafortune, 2006)). The language *generated* by $G = (X, E, f, \Gamma, x_0, X_m)$ is

$$\mathcal{L}(G) := \{s \in E^* : f(x_0, s) \text{ is defined}\}.$$

The language *marked* by G is

$$\mathcal{L}_m(G) := \{s \in \mathcal{L}(G) : f(x_0, s) \in X_m\}.$$

E^* denotes the Kleene-closure of E and is the set of *all* finite strings of elements of E including the empty string ε.

Figure 2.3 shows an automaton modeling a valve. It has the two states VC for valve closed and VO for valve open. When the automaton is in its initial state VC and then changes to the state VO the event *open_valve* is produced. Hence the language of the system 'valve' can be produced with the automaton performing a state trajectory. In the automaton in figure 2.3 state VC is marked. Hence, each marked event trace $s \in \mathcal{L}_m(G)$ must end in this state.

A common strategy to model a system is to use different automata G_i to represent the system components and to combine these component models to a global system model. In (Cassandras and Lafortune, 2006) different composition approaches like parallel or synchronous composition to combine several automata are given. Applying these composition algorithms it is possible to build a complex system model based on relatively simple component models.

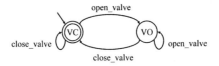

Figure 2.3: Automaton modeling a valve

2.2.3 Petri Nets

A second important modeling formalism for DES are Petri nets. They are especially suited for modeling systems with a high degree of concurrency in a compact manner. Like for automata, there exist many definitions of Petri nets and their according evolution rules. An example that is often referred to in literature is the labeled Petri net. The necessary definitions are taken from (Cassandras and Lafortune, 2006):

Definition 5 (Marked Petri net, (Cassandras and Lafortune, 2006)). A Petri net graph PN is a weighted bipartite graph

$$PN = (P, T, A, w, x)$$

where P is the finite set of places, T is the finite set of transitions, $A \subseteq (P \times T) \cup (T \times P)$ is the set of arcs from places to transitions and from transition to places in the graph, $w : A \rightarrow \{1, 2, 3, \ldots\}$ is the weight function on the arcs and x is a marking of the set of places P. $\mathbf{x} = [x(p_1), x(p_2), \ldots, x(p_n)] \in \mathbb{N}^n$ is the row vector associated with x and contains the number of tokens in each place.

The marked Petri net only represents the structure of the Petri net. In order to create a language, the Petri net must have well defined dynamics which are given in the next definition:

Definition 6 (Petri net dynamics, (Cassandras and Lafortune, 2006)). The state transition function $f : \mathbb{N} \times T \rightarrow \mathbb{N}^n$ of Petri net (P, T, A, w, x) is defined for transition $t_j \in T$ if and only if

$$x(p_i) \geq w(p_i, t_j) \text{ for all } p_i \in I(t_j)$$

with $I(t_j)$ representing the set of input places of transition t_j. If $f(\mathbf{x}, t_j)$ is defined, then we set $\mathbf{x}' = f(\mathbf{x}, t_j)$, where

$$x'(p_i) = x(p_i) - w(p_i, t_j) + w(t_j, p_i), \quad i = 1, \ldots, n.$$

With this definition, a transition function is only defined if all of the input places have at least as many tokens as the weight function delivers for the arc connecting the input place and the transition. In this case, the transition is enabled. If an enabled transition t_j fires, the input place p_i loses as many tokens as the weight function value is of the arc from p_i to t_j. If p_i is an output place of t_j, it gains as many tokens as the weight function value of the arc from t_j to p_i. After the definition of the Petri net dynamics it is possible to determine a token flow in the net. In order to create a language when performing the token flow according to definition 6 it is necessary to label the transitions in the Petri net with events:

Definition 7 (Labeled Petri net, (Cassandras and Lafortune, 2006)). A labeled Petri net N is an eight-tuple

$$N = (P, T, A, w, E, \ell, \mathbf{x_0}, \mathbf{X_m})$$

where P, T, A, w are given in definition 5, E is the event set for transition labeling, $\ell : T \to E$ is the transition labeling function, $\mathbf{x_0} \in \mathbb{N}^n$ is the initial state of the net (i.e. the number of initial tokens in each place) and $\mathbf{X_m} \subseteq \mathbb{N}^n$ is the set of marked states (i.e. final number of tokens in each place) of the net.

The notion 'marked states' is analogous to the marked states in definition 3. The language generated by the labeled Petri net is built by event traces that occur during a token flow starting in the initial setting $\mathbf{x_0}$. If the flow ends in one of the marked states $\mathbf{x_m} \in \mathbf{X_m}$, the trace is part of the marked language:

Definition 8 (Languages generated and marked, (Cassandras and Lafortune, 2006)). The language generated by labeled Petri net $N = (P, T, A, w, E, \ell, \mathbf{x_0}, \mathbf{X_m})$ is

$$\mathcal{L}(N) := \{\ell(s) \in E^* : s \in T^* \text{ and } f(\mathbf{x_0}, s) \text{ is defined}\}.$$

The language marked by N is

$$\mathcal{L}_m(N) := \{\ell(s) \in \mathcal{L} : s \in T^* \text{ and } f(\mathbf{x_0}, s) \in \mathbf{X_m}\}.$$

An example of a Petri net modeling a queuing system is given in figure 2.4. For the Petri net in the figure, the weight function is assumed to have the value 1 for each arc. In figure 2.4 the net is shown in two situations. On the left, the queue is empty and waits for costumers. On the right, the Petri net is shown after some events have occurred. The Petri net contains three events a (customer arrives), s (service starts) and c (service completes and customer departs), three transitions and three places. Place Q represents the queue, place I represents the condition 'server idle' and place B a busy server. Event a can occur spontaneously like a customer. When a appears a token is added to place Q to represent a customer in the queue. If the server is idle ($x(I) = 1$) the transition with the label s can fire which adds a token to place B and removes a token from Q. The marking of the Petri net on the right side of figure 2.4 can be obtained after the event trace $\{a, s, a, a, c, s, a\}$ and represents two customers waiting in the queue and one being treated by the server. It can be seen that the Petri net compactly expresses the fact that event a can occur at each time *concurrently* with the other two events since neither s nor c has an influence on a.

Automata and Petri nets are closely related. If the number of reachable Petri net markings is finite, an equivalent automaton can be created by a the construction of a reachability graph (Murata, 1989).

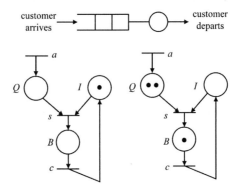

Figure 2.4: Sample Petri net

2.3 Diagnosis of Discrete Event Systems

2.3.1 Fault diagnosis in industrial systems

Introduction

Industrial systems have the purpose to perform a given production task in a given time and at given costs. Duration and costs of a production task are an important factor for the competitiveness of a manufacturing company. Only if both factors are minimized, the industrial facility reaches a high degree of productivity which has a positive effect on the competitiveness of a company. Duration and costs of a production process are increased if faults prevent the system from running in a normal mode. Faults often lead to so called 'downtimes' during which the system is not running and needs to be repaired. In order to minimize system downtimes, it is necessary to quickly get precise information about a fault after it occurred. The process of determining this information is referred to as *fault diagnosis* and consists of the following three steps (Chen and Patton, 1998):

- *Fault detection* is a decision – either that something is wrong or that everything works under normal conditions

- *Fault isolation* is the determination of the location of the fault (e.g. which sensor or actuator is faulty)

- *Fault identification* determines the size and the type or nature of the fault.

The approaches developed in this work are focused on fault detection and fault isolation. After a fault has been detected and isolated (i.e. the faulty component has been determined), the fault identification step will be left to the user and his technical experience. Since in large industrial facilities, faulty components are most often directly replaced instead of being repaired on-site, the exclusion of fault identification does not represent an important limitation.

In the field of fault diagnosis for technical systems, the two terms 'fault' and 'failure' are often used to describe a malfunctioning. In order to distinguish the two notions, two definitions taken from (ANSI/IEEE100, 1997) are given. A component fault is defined as follows:

> *Fault:* A physical condition that causes a device, a component, or an element to fail to perform in a required manner, for example, a short-circuit, a broken wire, an intermittent connection.

A fault does not necessarily lead to an unavailable system. In case of *fault tolerant* systems it is possible to implement a 'work around' using inherent redundancy to prevent the system from failing (Blanke et al., 2006). The term to describe that the system operation is no longer possible is 'failure' (ANSI/IEEE100, 1997):

> A *failure* is the termination of the ability of an item to perform a required function.

Hence, it is possible to avoid system failures by the early detection of faults.

Diagnosis approaches for technical systems are based on the assumption that a fault leads to some measurable or visible abnormality in a system parameter. This abnormality is referred to as *fault symptom* (ANSI/IEEE100, 1997).

Diagnosis principles

Fault diagnosis systems analyze measurements of a considered system and apply specific algorithms to deliver information about faults in the system (Korbicz et al., 2004). Various methods have been proposed by the scientific community and implemented in industrial applications. Most of the approaches can be classified in three groups (Papadopoulos and McDermid, 2001): Rule-based expert systems, data-driven approaches and model-based approaches.

Rule-based expert systems use specific system knowledge of an expert to perform the diagnosis task. For this reason it is necessary to formalize the available system knowledge to make it accessible for diagnosis algorithms. A possible formalization are IF-THEN-ELSE rules which can be evaluated using forward or backward chaining of rules (Papadopoulos and McDermid, 2001). In this process one or more rules are triggered by some deviation of a system parameter. Triggered rules can again trigger other rules in a chain. If no more rules can be triggered the resulting information (so-called *facts*) is supposed to contain the necessary data about the fault. Other possibilities to formalize the expert knowledge are summarized in (Venkatasubramanian et al., 2003) and (Isermann, 2006). Expert systems have successfully been applied to processes that can be characterized by a small number of rules. Larger systems often lead to inconsistencies and incompleteness among the rules which degrades the quality of the diagnosis system (Papadopoulos and McDermid, 2001). Another disadvantage is that expert systems need a considerable effort to apply them to new systems due to their high degree of individuality.

Data-driven approaches for fault diagnosis rely on the analysis of measured system data and the extraction of special features. An example for this class of approaches is trend analysis (Dash and Venkatasubramanian, 2000). For a given set of faults, expected trends in sensor measurements must be stored in a knowledge data base. (Dash and Venkatasubramanian, 2000) propose seven general trend types (so called primitives like a monotonic increasing or a constant signal value) to classify different faults. During online fault diagnosis sensor measurements are analyzed using appropriate filters to identify significant trends that correspond to a given setting of primitives. During this process, fault detection time and robustness of the method must be balanced: The analyzed measurement must be long enough to avoid misinterpreting 'normal' noise as a trend but short enough to detect real trends indicating a fault as quickly as required. Instead of using simple trend types like the primitives of (Dash and Venkatasubramanian, 2000) it is also possible to use more complex data characterizations like patterns (Papadopoulos and McDermid, 2001).

The third class of diagnosis approaches is *model-based*. The idea of model-based methods is to compare the observed system behavior with the expected behavior defined by a system model. Figure 2.5 shows this principle. A possible fault detection policy within this framework is to detect if the observed system output and the predicted system output differ significantly. Model-based methods can be divided in two subclasses (Isermann, 2006): Methods using models of the nominal system behavior only and methods using explicit fault models.

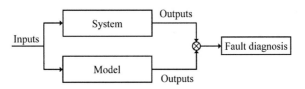

Figure 2.5: Principle of model-based diagnosis

Using models of the nominal system behavior only, it is possible to detect faults if they lead to system behavior that cannot be reproduced by the model. In the case of continuous systems, fault diagnosis is often performed using special *residuals* where measured and modeled signals are systematically compared (Isermann, 2006). If given thresholds are exceeded, a fault is detected. Considering the precise evolution of the residuals it is also often possible to isolate a fault. Another approach for working with fault-free models is given by (Reiter, 1987) in the field of artificial intelligence: Like in the case of continuous systems, a fault is detected if the measured behavior cannot be reproduced by the model. To determine a set of fault candidates, the minimal set of components is calculated that has to be considered as faulty to explain the deviation.

Especially if detailed system knowledge about the effects of faults on the system behavior is available, it is an established approach to explicitly model the faulty behavior. In this case, online fault diagnosis is performed by comparing measured signals with different scenarios of the model output. If in one scenario the observation is consistent with an explicitly modeled faulty behavior, it is possible to precisely identify the according fault. Examples for this approach are given in (Sampath et al., 1996) for DES or (Isermann, 2006) for continuous systems.

The decision for an appropriate diagnosis principle for a given application may always be influenced by personal preferences of the responsible engineer. However, there are some important arguments leading to *model-based* approaches as the preferable principle for closed-loop DES. Since industrial closed-loop DES are usually large systems, the application of *rule-based expert systems* is not possible. Capturing the necessary knowledge for diagnosis with consistent rules needs a considerable effort which may often be too expensive. *Data-driven approaches* mainly rely on the analysis of trends or patterns in the measured system data. For continuous signals for example, it is possible to define trends by an analysis of the signal amplitude. In event sequences emitted by closed-loop DES such an analysis is not possible. Although there exist approaches to determine event frequency or event patterns (Ghallab, 1996) they are proposed with a special background: their aim is to process abstract high-level data like the occurrence of alarm events in SCADA (Supervisory Control and Data Acquisition) or in telecommunication systems (Cordier and Dousson, 2000). In these applications there already exist diagnosis entities creating a huge amount of status or alarm messages that have to be preprocessed and filtered before given to a human system operator. Hence, data-driven approaches are not adapted for working on the external behavior of closed-loop DES which is defined by sensor readings and actuator settings. *Model-based* approaches in contrast are suitable to represent the external closed-loop DES behavior since appropriate modeling formalisms exist (see sections 2.2.2 and 2.2.3). The advantage of model-based methods in DES is their notion of *state*. Since closed-loop DES are dynamical systems, the decision if a given event must be considered as a fault often depends on the current system state. Using DES-models like automata or Petri nets, the system state can be efficiently represented which allows tolerating an event in one system state and detecting a fault upon observation of the same event in another state. Another advantage of model-based methods is that once the models are obtained (manually or by identification) they can be reused for other purposes like formal verification (Machado et al., 2006) or re-engineering (Frey and Younis, 2004) to further improve system dependability.

2.3.2 Model-based diagnosis with fault models

Diagnosis with observer automata

One of the most prominent model-based diagnosis methods for DES is the diagnoser approach of (Sampath et al., 1996). The aim of this approach is to decide if an *unobservable* event (usually a fault) has occurred only taking into account strings consisting of *observable* events. The first step of the method is to build models of the system components and of the algorithm controlling the system behavior. The models are built using finite automata (definition 3). During the modeling process the normal behavior and the behavior in case of given faults must be explicitly included in each component model. In the approach it is assumed that the event set E of a system can be divided into observable (E_o) and unobservable events (E_u) leading to $E = E_o \cup E_u$.

As an example, a simple system consisting of a valve, a controller and a flow sensor is considered. In the example it is assumed that under normal (i.e. fault-free) conditions a liquid flows through a pipe connected to the valve if the valve has been opened by the controller. Figure 2.6 shows the according automata models of valve and controller. The *normal* behavior of the valve is the same as in the example in figure 2.3. Additionally, there are two states representing the valve after the occurrence of a fault. The state SC represents the valve under a 'stuck close' fault and state SO models 'stuck open'. For the controller no faults are modeled. Controller and valve are synchronized using the *observable* events *open_v* and *close_v*. The occurrence of a fault in the valve is represented by one of the *unobservable* events *fail_close* or *fail_open*. In the example it is assumed that the stuck open fault can only occur when the valve is open whereas the stuck close fault can occur when the valve is open and closed.

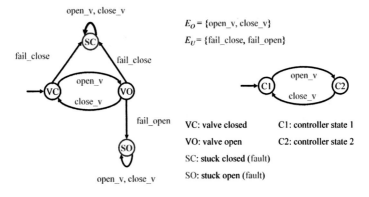

Figure 2.6: Models of valve and controller

After the component models have been built, the global system model can be obtained by parallel composition. In (Cassandras and Lafortune, 2006) the parallel composition of two automata G_1 and G_2 of the type given in definition 3 is defined as follows:

Definition 9 (Parallel composition). The *parallel composition* of G_1 and G_2 is the automaton

$$G_1||G_2 := Ac(X_1 \times X_2, E_1 \cup E_2, f, \Gamma_{1||2}, (x_{01}, x_{02}), X_{m1} \times X_{m2})$$

where

$$f((x_1, x_2), e) := \begin{cases} (f_1(x_1, e), f_2(x_2, e)) & \text{if } e \in \Gamma_1(x_1) \cap \Gamma_2(x_2) \\ (f_1(x_1, e), x_2) & \text{if } e \in \Gamma_1(x_1) \backslash E_2 \\ (x_1, f_2(x_2, e)) & \text{if } e \in \Gamma_2(x_2) \backslash E_1 \\ \text{undefined} & \text{otherwise} \end{cases}$$

and thus

$$\Gamma_{1||2}(x_1, x_2) = (\Gamma_1(x_1) \cap \Gamma_2(x_2)) \cup (\Gamma_1(x_1) \backslash E_2) \cup (\Gamma_2(x_2) \backslash E_1)$$

The Ac operation delivers the *accessible* part of the automaton, i.e. only states reachable by a state trajectory starting in the initial state are part of $Ac(G)$.

After the parallel composition procedure, common events of two automata (events that belong to E_1 and E_2) can only occur if the two automata both execute the according event simultaneously. Events belonging only to one of the two automata are not synchronized and can thus occur whenever it is possible.

Figure 2.7 shows the result of the parallel composition of valve and controller automata from figure 2.6. Since both automata contain the events *open_v* and *close_v*, in the parallel composition these events can only occur if they are enabled in both of the underlying automata. The states of the parallel composition in figure 2.7 also contain additional information of the flow sensor in the valve-controller example. This sensor indicates if there is flow in the pipe that is connected to the valve. In the approach of (Sampath et al., 1996) it is necessary to add the sensor information to the parallel composition states *manually*.

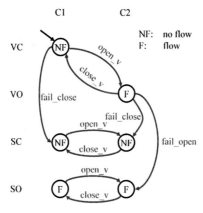

Figure 2.7: Parallel composition of valve and controller automata

The next step of the (Sampath et al., 1996) approach is to include the sensor information into the events of the parallel composition. Each *observable* event e_o is replaced by $< e_o, h(x') >$ where $h(x')$ denotes the sensor value of the following state that is reached when e_o is produced. For each *unobservable* event at a transition it is checked if the sensor has the same value in the source state x_1 and the target state x_2. In this case, the unobservable event e_u between these two states is replaced by $< e_u >$. If the sensor value differs between source and target state, the according transition is removed and an additional state is added to the automaton. This state can be reached from the source state x_1 producing the unobservable event $< e_u >$. The new state can be left to the target state x_2 using a transition with the newly created *observable* event $< h(x_1) \rightarrow h(x_2) >$. On the left side of figure 2.8 the parallel composition after the inclusion of the sensor information in the events is shown. The black state has been added to the automaton following the procedure described above.

The automaton still contains the unobservable events $fail\_close$ and $fail\_open$ modeling the occurrence of a fault. Since the aim of the (Sampath et al., 1996) approach is to analyze only *observable* events to find faults in the system, a special diagnoser automaton is derived. The diagnoser depicted on the right side of figure 2.8 is constructed according to the procedure given in (Sampath et al., 1996). It only contains observable events and estimates the current system states. In each of its states, a set of possible current model states is given which can be reached by the event sequence observed so far. The according state number is given with a so called 'fault label'. For each state in the system model in the left side in figure 2.8, it must be decided if this state is a normal state (label N) or if it is a fault state (label Fi). In the example, $F1$ represents the fault 'valve stuck close' and $F2$ represents 'valve stuck open'. In the initial diagnoser state e.g. two model states are given as possible state estimates. Both states x_1 and x_4 can be the actual system state if no event has yet been observed: The first possibility is that no event occurred in the system. In this case x_1 is the actual (and initial) state. The other possibility is that the *unobservable* event $< fail\_close >$ has occurred which led to a transition from x_1 to x_4.

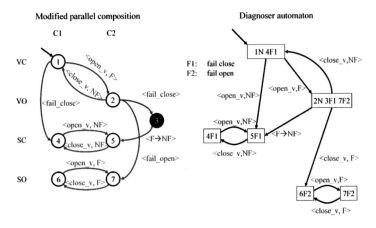

Figure 2.8: Modified parallel composition and diagnoser automaton

One of the main features of the diagnoser approach of (Sampath et al., 1996) is the possibility to determine if a given fault is diagnosable with the derived diagnoser. In (Cassandras and Lafortune, 2006) diagnosability is defined as follows:

Definition 10 (Diagnosability according to (Cassandras and Lafortune, 2006)). Unobservable event e_d is *not diagnosable* in live language $\mathcal{L}(G)$ if there exist two strings s_N and s_Y in $\mathcal{L}(G)$ that satisfy the following conditions: (i) s_Y contains e_d and s_N does not; (ii) s_Y is of arbitrary long length after e_d; and (iii) $P(s_N) = P(s_Y)$. When no such pair of strings exists, e_d is said to be *diagnosable* in $\mathcal{L}(G)$. $P(s_N)$ denotes the projection of the string s_N to the alphabet of observable events E_o. This operation removes each event in string s_N which is not part of E_o.

The notion 'live language' refers to a language without terminating strings. An automaton without deadlock states is capable of generating a live language. Definition 10 means that a fault which leads to the unobservable event e_d is only diagnosable if each observable event sequence following e_d can eventually be distinguished from any other event sequence of the system language. Hence, the observable part of the system behavior after fault e_d must be distinguishable from any other string that can occur without e_d. In (Sampath et al., 1996) an algorithm is given to test if a given set of faults is diagnosable with the derived diagnoser.

The diagnoser approach of (Sampath et al., 1996) has been modified and improved in many ways. In (Debouk et al., 2000) a *decentralized* architecture using a set of diagnosers to observe a system is proposed. The principle of the architecture is depicted in figure 2.9. In the decentralized approach it is assumed that the global system information is distributed at several sites. At each site only a subset of events can be observed. In the distributed diagnoser framework, for each site an own diagnoser has to be constructed according to the procedure explained above. With this approach it is possible to limit the size of the automata used for online-fault diagnosis since the local diagnosers are typically much smaller than a diagnoser for the complete system. Since each local diagnoser has only limited access to the necessary system information, it can be necessary to implement communication between the diagnosers or communication with a coordinator.

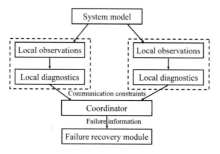

Figure 2.9: Distributed diagnoser architecture (Debouk et al., 2000)

A second major improvement of the diagnoser approach of (Sampath et al., 1996) is to include information of event *timing* into the system models. With this approach it is possible to find faults leading to a changed timed system behavior or to deadlocks. In (Hashtrudi Zad et al., 2005) a theoretical framework for timed diagnosis is introduced. The main idea is to represent time by a repeatedly occurring event, the so-called 'tick'. With this special event it is possible to include e.g. the expected number of 'ticks' between two regular events in the component models. The enhanced system models can then be processed like in the approach of (Sampath et al., 1996) to construct the diagnoser.

A combination of the distributed and the timed approach summarized above is proposed in (Philippot et al., 2007) and (Sayed-Mouchaweh et al., 2008). This approach is especially well suited for diagnosis in manufacturing systems which are a special class of closed-loop discrete event systems introduced in section 2.1. Like in the classical diagnoser approach from (Sampath et al., 1996) the first step of the method is to build models for the plant components. As a difference to the classical diagnoser approach, only the fault-free system behavior must be modeled. In (Philippot, 2006) it is explained how so called 'plant elements' can be obtained in a systematic way by composing models of the actuator and of the sensor behavior of each component. The component models have to be synchronized with the control algorithm to represent the desired controlled behavior. In (Philippot et al., 2005) a procedure to extract the relevant part of the control algorithm for each component model from the control algorithm is proposed. A formal description of the control algorithm must be available in form of a GRAFCET (GRAphe Fonctionnel de Commande Etapes/Transitions) which is a special specification language for automation systems (IEC, 2002). After the composition of component models and the according parts of the control algorithm to automata models, timing information of the event occurrences must be added. For each state in the controlled component models, valid time windows for the occurrence of the following events must be determined. The result of this step are timed controlled component models of the normal system behavior. These models are the basis for the construction of the distributed diagnosers. Following special rules, some faulty behavior is added to the controlled component models. Based on these enhanced models, diagnosers are derived which monitor the according component during system operation. In figure 2.10 a part of such a diagnoser is depicted. Even without a formal definition of the model semantics its working principles can be seen.

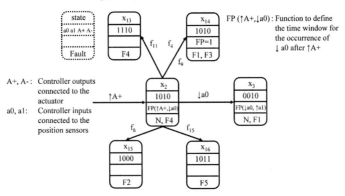

Figure 2.10: Part of a diagnoser from (Philippot, 2006)

The depicted diagnoser is derived from a model of a double acting cylinder with two positions (position sensor $a0$ for the initial position and position sensor $a1$ for the extended position). In each diagnoser state, the controller input/output values of the cylinder sensors and actuators are given (second line). In state x_2 the input/output setting is 1010 which means that the cylinder is in its initial position (first value of the setting is 1 which means that the according position sensor is activated) and has been ordered to extend (third value of the setting is 1 which means that the according actuator has been started by event $\uparrow A+$). The next expected action is the cylinder leaving its home position which leads to event $\downarrow a0$. In the third line of diagnoser state x_2 a time window for the occurrence of $\downarrow a0$ after $\uparrow A+$ is given. If the maximum waiting time is exceeded, the diagnoser goes to state x_{14} which allows concluding that one of the faults with labels $F3$ or $F4$ must have occurred (defect at sensor $a0$ of actuator defect). Since for each component an own diagnoser monitors the system, a global coordinator decides if a fault message must be issued to the system operator.

Diagnosis with state estimation algorithms

A common characteristic of the methods sketched in the former section is their working principle based on observer automata. For diagnosis it is necessary to have a current state in the diagnoser during system monitoring since the diagnoser states contain the state estimation for the considered system. Hence, system and diagnosers must be started synchronously which can be difficult. To overcome this problem, in (Neidig and Lunze, 2006) a state observation algorithm directly working on the system model is proposed. It starts with a given initial state estimate. If no information about the system state is available, it is possible to start with the complete state space of the system model as initial state estimation. The currently measured system data must be given to the algorithm. Based on the current state estimation it determines which model states can be reached with the measured system data. These states become the new estimate. If models containing the fault-free and some faulty system behavior are to be used, for each model state it is to decide a priori if it represents a faulty or a fault-free behavior. Based on this information it is possible to decide if the current state estimate contains faulty states. Hence, the state observation algorithm can be used as a diagnosis procedure which replaces the diagnoser introduced in the former section. Since the state estimation algorithm derives the current state estimate 'on the fly', it is more demanding in terms of calculation time compared to evaluation of the diagnoser automaton. As the diagnoser can be constructed offline, it is possible to perform more complex calculations for its construction.

Like the classical diagnoser approach, the method using a state estimation algorithm has been modified and improved in many ways. In (Supavatanakul et al., 2006) it is shown that this approach can be adapted to timed models. For this purpose the state estimation algorithm also considers the time intervals between measured system signals and integrates this information in the state estimation procedure. In (Neidig and Lunze, 2006) an approach for coordinated diagnosis of automata networks similar to the distributed approaches for the diagnoser method is proposed. Since in some technical applications it is possible to give an initial state estimation with probability values for each state in the estimate, the method has also been extended to stochastic systems (Neidig and Lunze, 2005). For this adaptation a stochastic model of the considered system is necessary.

Similar to the diagnoser approach and its derivatives, it is possible to decide if a given fault can be diagnosed with the available models (diagnosability) using the state estimation approach.

Model-based diagnosis with Petri nets

The diagnosis approaches presented so far are working with different forms of automata to model the considered system. The system is usually modeled by composing component models represented by automata. During this process, model building for large systems often leads to the state space explosion problem which is the result of the numerous combined component behaviors captured within a composed model. An approach to avoid the state space explosion problem is to use Petri nets which are able to represent the system state in a distributed way. An overview of fault diagnosis approaches using Petri nets is given in (Pia Fanti and Seatzu, 2008). In Petri net approaches, the models used for system monitoring (Petri net diagnosers) inherit the structure of the underlying system. Some places or transitions in the system model are assumed to be non-measurable. For non-measurable places this means that the number of tokens is not known. Non-measurable transitions on the other hand can fire without being noticed. The diagnosis problem consists in determining the distribution of tokens in the Petri net over measurable and non-measurable places (or the according firing sequence of non-measurable transitions) only taking into account measurable system information. Like in the case of diagnosis with automata, Petri net approaches for monolithic and distributed models exist.

In (Genc and Lafortune, 2003) a Petri net approach assuming that all places of the system model are non-measurable is proposed. It is based on the class of labeled Petri nets introduced in definition 7 with an event set consisting of observable and unobservable (fault) events. Events are associated with transitions. The diagnosis problem is to determine which unobservable events can have occurred based on an analysis of observable events. The diagnoser Petri net in this approach inherits the *complete* structure of the Petri net modeling the system (including the structure of the faulty behavior). Meanwhile the dynamics of the system are defined by observable and unobservable events, the Petri net diagnoser evolves only based on observable events. Based on the observable events it determines which states could have been reached in the system if some unobservable events occurred. Figure 2.11 shows a system model and its Petri net diagnoser. Transitions with unobservable events are gray. Although the Petri net diagnoser in the right part of the figure has the same structure as the system model on the left, its state is defined in a different way. Each diagnoser state represents several possible system markings each represented by the distribution of another token symbol. The marking with the •-symbol (p_1 and p_2) represents the system state like depicted on the left side of the figure. Starting from this system state it is possible that the unobservable event g occurs leading to the marking p_3 and p_2. This system state is represented by the ■-symbols. If the unobservable event g occurred twice, the resulting marking is p_5 and p_2 which is represented by the distribution of the +-symbol in the diagnoser state. Using this principle, a diagnoser Petri net state contains each possible system state that is reachable by the emitted series of observable events and arbitrary occurrences of unobservable events. The Petri net diagnoser can be calculated offline. If the underlying Petri net model is very large and complex, the calculation of the diagnoser leads to the state space explosion problem since each possible state of the underlying Petri net must be enumerated. Like in the case of automaton diagnosers, this problem is addressed using a distributed approach with communicating Petri diagnosers for different system components (Genc and Lafortune, 2007).

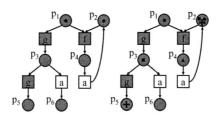

Figure 2.11: Petri net diagnoser according to (Genc and Lafortune, 2003)

The calculation of the Petri net diagnosers can result in very large models with considerable memory demands which can be an obstacle for online application. Another drawback common to the diagnoser approach using observer automata is that Petri net diagnoser and system must be started synchronously to assure an accurate initial state estimation. Instead of the offline calculation of the Petri net diagnoser, in (Dotoli et al., 2009) it is proposed to determine the possible marking online solving an integer linear programming problem (ILP). The ILP is formulated using constraints for firing sequences in the given net structure that must be fulfilled to reproduce the online observation with the model. These constraints are formulated with matrix equations using e.g. reformulations of the state transition equation $\mathbf{x}' = f(\mathbf{x}, t_j)$ (see definition 6 of the Petri net dynamics). For each fault considered in the model an ILP has to be solved maximizing the occurrence of the concerning fault event under the constraint that the trajectory is valid according to the Petri net structure and dynamics and produces the observed event sequences. If the ILP admits a solution > 0 for a given fault, the occurrence of this fault is a possible explanation for the observed event sequence: a firing sequence of observable and unobservable events exists reproducing the observed sequence and containing the considered fault. If non of the ILPs belonging to the faults included in the Petri net model leads to a solution > 0, it is concluded that the behavior is normal. The approach of (Dotoli et al., 2009) reduces the amount of necessary memory to store the diagnoser model at cost of increased calculation time. Since for solving ILPs there does not exist a polynomial algorithm, the model structure has to keep to some constraints (like representing an acyclic state machine) to guarantee that the ILPs can be solved fast enough to be applied online. However, for large systems this approach does not admit solutions in a short time such that it could be applied online.

2.3.3 Model-based diagnosis with fault-free models

After the presentation of the major works on diagnosis with fault models for DES, in this chapter methods on diagnosis with fault-free models are investigated. In (Reiter, 1987) a very general theory for diagnosis with a fault-free system description is given. The method is situated in the field of Artificial Intelligence (AI) and is working on an abstract system description which can be given in different kinds of logic. A system is described as set of components interacting according to a system description:

Definition 11 (System (Reiter, 1987)). A *system* is a pair $(SD, COMPONENTS)$ where:

1. SD, the *system description*, is a set of first order sentences;

2. $COMPONENTS$, the *system components*, is a finite set of constants.

The system description can be given in any suitable logic like first-order, dynamic or temporal logic. For fault detection it is checked if the current system observation OBS is consistent with the system description SD and with the assumption that each component is working normally. To denote that a component c_i is not abnormal (AB) the formulation $\overline{AB}(c_i)|c_i \in COMPONENTS$ is used. Hence, it is checked if

$$SD \cup (\overline{AB}(c_1), \ldots, \overline{AB}(c_n)) \cup OBS$$

is consistent. If this consistency check fails, a fault is detected. This procedure is similar to the comparison of observed and modeled behavior in model-based diagnosis approaches. In the approach of (Reiter, 1987), diagnosis is considered as a *conjecture* that certain system components are faulty and some are normal. The aim of diagnosis is to estimate the set of faulty components as precisely as possible. This is done using the the Principle of Parsimony:

> *The Principle of Parsimony.* A diagnosis is a conjecture that some minimal set of components are faulty.

This heuristic expresses the idea that there is usually a higher probability for a given component to be fault-free than to be faulty. Hence, in case of fault detection, it is reasonable to determine a *minimal* set of components which have to be assumed abnormal ($\overline{AB}(c_i)$) to restore consistency of observation and system description. This consideration results in the following definition for the term diagnosis:

Definition 12 (Diagnosis (Reiter, 1987)). A diagnosis for $(SD, COMPONENTS, OBS)$ is a minimal set $\Delta \subseteq COMPONENTS$ such that

$$SD \cup OBS \cup \{AB(c)|c \in \Delta\} \cup \{\overline{AB}(c)|c \in COMPONENTS \backslash \Delta\}$$

is consistent.

Several approaches to determine the minimal set Δ containing the diagnosed components exist. A straightforward approach is to systematically generate all subsets $\Delta \subseteq COMPONENTS$, starting with minimal set cardinality first, and testing the consistency of system description, observation and assumed malfunctioning components. In (Reiter, 1987) the diagnosis approach is applied to digital circuits using first order logic.

The approach of (Reiter, 1987) has not been developed for DES although they can also be represented using logic descriptions. An approach for model-based diagnosis based on fault-free models explicitly developed for DES is given in (Pandalai and Holloway, 2000). It has been proposed for diagnosis in manufacturing systems with discrete sensors and actuators. The main idea of the method is to monitor the timing and sequencing of events generated by the considered system. The nominal (fault-free) system behavior is modeled using *condition templates*. A condition template is determined by a set of rules starting from an initial triggering event: based on the initial event occurrence, a set of *expectations* is determined that contains possible following events with an appropriate timing interval. If one of the expected events occurs within a valid time interval, new expectations are created according to the rules given in the condition template. An expectation is given in the form (t, e, C, w) where t is a time, e an event, C a consequence and w a tag (label of the expectation). t is the time stamp when the triggering event e occurs. The consequence C is a pair (e', τ), where e' is an event and τ is a time interval.

During system monitoring a set of expectations is maintained online. Following the rules in the condition template, expectations are added or removed from the expectation set after the occurrence of the according events. Two scenarios for fault detection exist in this framework: Either non of the events in an expectation occurred within the specified time periods, or an event occurs that was not expected by any of the active expectations. In the sense of (Reiter, 1987) the event which cannot be explained by one of the expectations representing the system description is considered as fault event. Assuming the component related with the suspected event to be faulty restores consistency of observation and system description. In figure 2.12 an example of system monitoring using condition templates is given. Two expectations are depicted: $(2, e_1, \{(e_2, [2, 4]), (e_3, [6, 8])\}, w_1)$ on the top and $(9, e_3, \{(e_5, [3, 5])\}, w_2)$ on the bottom. The first two elements in expectation w_1 indicate that event e_1 occurred at time $t = 2$. The gray shaded zones indicate the time intervals τ when the occurrence of e_2 or e_3 is expected during fault-free system behavior. The expectation is satisfied if e_2 appears at time interval $4 \leq t \leq 6$ or if e_3 appears at time interval $8 \leq t \leq 10$. If e_3 appears, the next expectation w_2 is triggered waiting for event e_5 at time interval $12 \leq t \leq 14$. In the example, the occurrence of event e_2 (dashed arc) is not expected by expectation w_1. Hence, it is inferred that e_2 occurred (to late) due to a fault.

Modeling the nominal system behavior with the framework from (Pandalai and Holloway, 2000) is not a standard tool in DES-theory. However, the authors present a method to systematically derive the condition templates from timed automata representing the nominal system behavior.

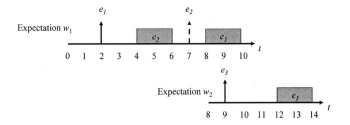

Figure 2.12: Graphical representation for template

2.3.4 Conclusion

In section 2.3.1 it has been explained that *model-based* approaches for diagnosis in closed-loop DES are preferable to *rule-based expert systems* or *data-driven methods*. Since model-based approaches derive information from the estimated system state, they allow a dynamic analysis of the external DES-behavior.

A common feature of diagnosis methods using explicit *fault models* is the diagnosability property: it is possible to check if a given (and modeled) fault can be diagnosed using the available models. Although this is an important advantage, it has to be kept in mind that the diagnosability property strongly depends on the quality of the system model. It only checks if a modeled fault can be diagnosed if the system behaves exactly like supposed in the model. For approaches using fault models, it is necessary to build models of the nominal and the faulty system behavior in form of automata or Petri nets. The determination of the nominal and the faulty behavior usually necessitates the deep knowledge of a system expert. This expert is not necessarily familiar with the modeling formalisms for DES. Even if it is possible to build the necessary models, none of the methods give guidelines of how to modify existing models in the case of failed diagnosability tests. Especially if distributed approaches are used to overcome the state space explosion problem, there are no guidelines of how to choose appropriate partial models. Some approaches propose model libraries with frequently used component models to facilitate model-building (Philippot et al., 2007). Although this may considerably reduce the effort to build component models, in most industrial systems there exist some non-standard components which have to be analyzed and modeled separately. Another important problem is the modeling process for the control algorithm. For this purpose it is necessary to have a formal description of the algorithm (e.g. as a GRAFCET). This is often not the case for existing industrial facilities where controller programs are rarely well documented. Another general problem for this class of methods is the fact that only faults explicitly considered in the model can be diagnosed. As a conclusion, methods working with fault-models are well adapted if diagnosability has to be guaranteed for some faults (e.g. safety related issues) and if there are sufficient financial means for detailed and exact model-building. To the best knowledge of the author, none of the diagnosis methods of this class has yet been successfully applied to a system at industrial scale.

In the second class of model-based methods working with *fault-free models*, the notion of diagnosability does not exist. Faults can be detected if they lead to a deviation from the modeled nominal behavior. It is not possible to guarantee that a given fault can be diagnosed using fault-free models. In general, fault diagnosis tends to be less precise using fault-free models since the models contain less knowledge. If the effect of a fault is not explicitly part of the model, the diagnosis information must be derived from the comparison of expected fault-free (modeled) and observed behavior. As explained in the approach of (Reiter, 1987), there may be several possible ways to explain this difference. An important advantage of approaches with fault-free models is that model-building is usually less laborious since only the fault-free behavior has to be represented. It is even possible to get the necessary fault-free models by identification. In the case of industrial closed-loop DES it is possible to observe the considered system during normal operation and to identify a DES-model based on the observed data. Although diagnosis with fault-free models has some disadvantages since diagnosability for given faults cannot be guaranteed, it offers a way to implement a diagnosis system without expensive model-building if identification methods are used. For many closed-loop DES at industrial scale it represents the only way to come to a diagnosis system since model-building would quickly exceed financial and organizational limits.

2.4 Identification of Discrete Event Systems

2.4.1 Limits of manual model-building and aims of identification

A key issue for model-based methods is the determination of appropriate models. Two main approaches are possible: *manual model-building* and *identification*. In case of *manual model-building*, the models are built by a human expert formalizing his knowledge about the causal structure and the behavior of the considered system in an appropriate mathematical form. *Identification* on the other hand refers to the systematic analysis of monitored system behavior and the algorithmic construction of a mathematical description which approximates the observed data. The core of an identification method is an identification algorithm.

Building a model *manually* necessitates knowledge of the considered system. Depending on the application, it is often possible to get an approximative model without a detailed analysis of the considered system. However, for the construction of exact models it is essential to have a deep understanding of the system to be modeled. An advantage of manual model-building is the possibility to gain deeper insights in the system structure and dynamics since inconsistencies in the system knowledge become visible when it is formalized in a model. Beyond the use in diagnosis systems, the improved system knowledge can be useful for many purposes like the development of control algorithms.

Many model-based techniques guarantee the correctness of certain results or their capability to perform a certain task. Model-checkers for formal verification for example guarantee that the complete modeled system behavior is analyzed to determine if given specifications hold (Bérard et al., 2001). A second example has been given in section 2.3.2: It has been shown that some diagnosis methods are able to determine if a given fault is diagnosable using the available models. If it is diagnosable, the methods guarantee its detection and correct indication. Since model-based methods cannot evaluate if a given model represents the considered system in an appropriate way, guaranteed features are only as good as the given models. It depends on the modeling engineer to make sure that the model is appropriate for the given purposes. An important question in this context is the correct level of abstraction. As an example for the consequences of this question we consider the different models of a valve in figure 2.13. The leftmost automaton has two states representing the valve in the opened and in the closed position. In this model it is assumed that the valve directly changes from opened to closed or vice versa when the according event occurs. The automaton in the middle of figure 2.13 has an additional state to represent the valve being between opened and closed. The rightmost model finally has four states. State B1 represents the valve moving from the closed position to the opened position whereas B2 models the valve moving from opened to closed. Each of the three models represents the valve behavior at a different level of abstraction. In the leftmost automaton intermediate states between valve open and valve close are not considered which may be sufficiently accurate for a given purpose. In the rightmost model, there is a more detailed representation of the valve behavior at cost of model size. If the more detailed model is e.g. used in the diagnoser approach, the parallel composition of plant models and controller algorithms becomes larger which also leads to a larger diagnoser automaton. The choice of an adapted level of abstraction is not obvious since the resulting model size has to balanced with the required model accuracy.

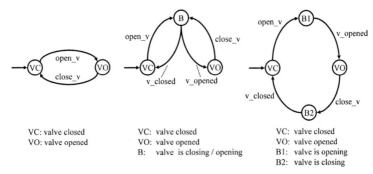

Figure 2.13: Different levels of abstractions for the valve model

Most model-based techniques require the model to be in a predefined form like differential equations in time-continuous systems or a certain type of automaton or Petri net in discrete event systems. A person with a sufficient degree of system knowledge is not necessarily familiar with the mathematical tools to build the model in the required form. Hence, it is often necessary to coordinate specialists from the domain of the considered system (e.g. mechanical engineers, chemists, etc.) with specialist of the model domain (e.g. control engineers) which poses an organizational challenge. The deployment of several specialists also leads to important economical charges which may lead to the decision that the application of model-based methods for a given scenario is too expensive.

For many applications it is possible to decrease the cost for model-building and even to increase model accuracy using *identification* methods. For identification it is necessary to observe the considered system for a certain time to get samples of its behavior. Based on the collected data, (van Schuppen, 2004) defines system identification as the selection of a model such that the observed data can be reproduced as accurately as necessary. The required accuracy is given by an approximation criterion.

Using identification methods to determine the models for model-based techniques has an important advantage: If it is possible to prove certain properties of an identification method, the resulting identified models have some guaranteed qualities. If an identification algorithm guarantees some degree of accuracy with respect to the collected system data, this leads to a more objective model evaluation than in the case of manually built models always depending on the competence of the modeling engineer.

Another advantage of identified models of closed-loop DES is that they directly represent the closed-loop of plant and controller. Manually built models usually only contain the *idealized* behavior of plant components and the control *algorithm* (e.g. in the diagnoser approach of (Sampath et al., 1996)). An identified model on the other hand represents the control algorithm *executed in the controller* as it leads to the observed system behavior. Special effects resulting from the controller hardware (e.g. delays of I/O cards) or the execution procedure of the algorithm (like cyclic program execution in industrial controllers) are thus often better represented using identified models since these effects are usually not considered in manual model building (Roth et al., 2010). Hence, identified models are often closer to the real system behavior than manually built ones.

In discrete event systems, the aim of identification is to build a model approximating the original system language L_{Orig}[1]. The basis for each identification approach is the observed language L_{Obs} which represents a (more or less complete) sample of the original system language $L_{Obs} \subseteq L_{Orig}$. It is supposed that the original system language is *fault-free*. The longer the observation horizon, the more likely a convergence of L_{Obs} to L_{Orig} is. Once a model has been identified on the basis of L_{Obs}, it creates a language which is referred to as the identified language L_{Ident}. If the original system language has been completely observed ($L_{Obs} = L_{Orig}$) and a given identification algorithm delivers a model which exactly reproduces the observed language ($L_{Ident} = L_{Obs}$) the aim of identification is perfectly met since $L_{Ident} = L_{Orig}$.

Figure 2.14 shows the relation of L_{Orig}, L_{Obs} and L_{Ident} in a general case. The area with the bold line represents L_{Orig} and contains the observed language L_{Obs} as a subset. In figure 2.14, the identified language is a superset of the observed language which is a reasonable minimal demand on an identified model used for diagnosis: it should be able to reproduce the already observed behavior of the considered system since it is fault-free.

---

[1]The precise definition of the term language often varies from one approach to another. Here, a language can be understood like introduced in definition 2. The considerations of this section also hold for the languages defined in the following chapters.

Usually, it cannot be guaranteed that an identified model only produces the observed language. Hence, L_{Ident} is a superset of L_{Obs} with some part of the set $L_{Ident}\backslash L_{Obs}$ possibly contained in the original language L_{Orig}. If an identified model contains an important amount of the language $L = L_{Ident}\cap L_{Orig}$, this increases the model accuracy. The identified language which is not part of the original system language is called exceeding language $L_{Exc} = L_{Ident}\backslash L_{Orig}$. The exceeding language generally decreases the model accuracy since each word of this language can be exhibited by the identified model but is not part of the original system language.

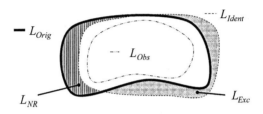

Figure 2.14: Relation of original, observed and identified languages

If it cannot be guaranteed that the original language is completely included in the identified language ($L_{Orig}\subseteq L_{Ident}$), it is also possible that some part of the original language cannot be reproduced by the identified model. The non-reproducible language L_{NR} represents the part of the original system language that has not been observed and that is not part of the identified behavior ($L_{NR} = L_{Orig}\backslash L_{Ident}$). An accurate model minimizes the two languages L_{NR} and L_{Exc} such that $L_{Ident} \approx L_{Orig}$.

For the decision if an identified model is appropriate for fault detection purposes, the exceeding language L_{Exc} and the non-reproducible language L_{NR} play a key role. If a system is diagnosed with the identified model and it exhibits a word $w \notin L_{Ident}$ a fault can be detected, since model and observation are not consistent (see section 2.3.3). In case that an exhibited word is part of the exceeding language ($w \in L_{Exc} = L_{Ident}\backslash L_{Orig}$), the word represents a fault ($w \notin L_{Orig}$) which cannot be detected since it is part of the exceeding behavior. In this case, the identified model erroneously contains a faulty word. To minimize the number of undetectable faults, it is necessary to minimize the exceeding language L_{Exc}.

If the system exhibits a word which is part of the non-reproducible language $w \in L_{NR} = L_{Orig}\backslash L_{Ident}$, a fault is detected, since $w \notin L_{Ident}$. This fault detection is a false alert since the exhibited word is part of the original system language L_{Orig}. A reduction of the number of such false alerts can be achieved by minimizing the non-reproducible language L_{NR}.

In the remainder of this chapter, different identification methods from the DES-domain are presented and it is discussed if the identified models meet the accuracy constraints of diagnosis systems.

2.4.2 Identification of finite automata

The first approaches for identification of DES have their origin in the computer science community and date back to the sixties and seventies of the last century. In (Klein, 2005) an overview of these methods is given and they are evaluated for identification of closed-loop DES. The approaches from this time can be divided in two main categories. Methods like (Biermann and Feldman, 1972), (Kella, 1971) or (Veelenturf, 1978) work on the basis of observed input and output sequences of the considered system and construct a Mealy or Moore automaton (Lee and Varaiya, 2002) to represent the observed data. The aim of these methods is to determine an automaton which is able to simulate[2] the observed language given by the captured input/output sequences. A frequent aim within this class of methods is to determine a minimal automaton which satisfies this constraint. The second class of methods like (Booth, 1967) aims at identifying completely specified automata using test sequences to excite the considered system. Completely specified automata are capable of producing an output sequence to each possible input sequence (Tornambe, 1996). For identification of industrial DES with physical components, the second class of methods is principally not adapted. The reason is that the application of various test input sequences is not possible for industrial closed-loop DES where valid system trajectories are predefined by the controller in the loop. Forcing the system to perform each possible trajectory quickly leads to critical situations.

In (Klein, 2005) it is shown that the methods proposed by the computer science community do not meet the requirements of models for online fault detection in industrial systems. These methods have not been designed for identification of systems with interaction of physical components and a controller like in the case of industrial closed-loop DES (see 2.1). Closed-loop DES exhibit a behavior without any input if the signals exchanged between controller and plant are considered as system output. From an external point of view they are event generators. The approaches summarized in (Klein, 2005) do not identify event generators but transducers which produce an output sequence upon the reception of an input sequence. Hence, they are not appropriate for modeling a closed-loop DES.

The early computer science methods summarized in (Klein, 2005) are based on various identification algorithms that process observed sequences and determine a model in a deterministic way. In (Baron et al., 2001b) and (Baron et al., 2001a), an identification approach that determines an automaton in a non-deterministic way is proposed. The key idea is to use genetic algorithms as an optimization technique to determine an optimal automaton structure which is able to reproduce the observed sequences. Genetic algorithms belong to the class of evolutionary, heuristic optimization approaches (Michalewisz and Fogel, 2000).

An important question when applying the evolutionary approaches to a technical optimization problem is how to represent possible solutions such that they can be treated like the genome in biological evolution. Figure 2.15 shows a Moore automaton from (Baron et al., 2001b) and a part of its representation as a 'genome'. The automaton is

---

[2]A language L_A simulates a language L_B if $L_A \supseteq L_B$ holds.

coded in a vector that can be divided in quintuples representing the different states of the automaton. The first entry of each quintuple contains the output of the state. The following four entries represent the states which are reached via a transition with a particular input (e.g. in the second following state entry of state 1 there is a 2, which means that from state 1, state 2 can be reached by input b (2)). Using genetic algorithms, two parent genomes have to be combined to determine new solutions. This procedure is not explicitly given in (Baron et al., 2001a) or (Baron et al., 2001b). Nevertheless, it is clear that two vectors like the one shown in figure 2.15 can be combined in different ways. One possibility is to randomly chose the quintuples describing the states of the new solution from one of the parents and to combine them to a new vector. The same approach is possible when choosing the precise values for the quintuple entries of state outputs or transitions leading to certain states. The concept of mutating a given solution can also be implemented following these ideas.

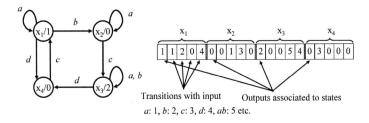

Figure 2.15: Example for the coding of a Moore automaton (Baron et al., 2001b)

For the assessment of a solution determined by recombination or mutation, it is necessary to determine its *fitness*. In (Baron et al., 2001b) an approach to determine the fitness by an analysis of the observed input/output sequence (only one such sequence in considered). Starting with the first I/O-pair, the length of the subsequence which is accepted by an automaton represented by a solution is determined. Solutions with higher fitness and thus longer accepted I/O sequences are more likely to be selected for reproduction which finally leads to a set of solutions with automata that are likely to accept an important part of the observed I/O sequence.

Although this approach is an interesting technique for identification of DES and shows the potential of optimization techniques in this field, the identified models are not well adapted for fault detection purposes. If the observed I/O sequence represents the fault-free behavior of the considered system, the identified (fault-free) model should contain this sequence to avoid false alerts (see section 2.4.1). Due to its non-deterministic nature, the approach of (Baron et al., 2001b) cannot guarantee that the observed sequence is completely accepted by the identified automata. It is also impossible to generally guarantee a certain accuracy of the identified model. In terms of section 2.4.1, it cannot be guaranteed that the exceeding language L_{Exc} and the non reproducible language L_{NR} are minimized. Hence, the models identified with the (Baron et al., 2001b) approach are not appropriate for fault detection purposes.

An automaton-based identification approach explicitly developed for fault diagnosis in DES is proposed in (Supavatanakul et al., 2006). The approach has been developed for diagnosis in continuous systems with a quantized I/O and state space. The amplitude of each continuous I/O signal is divided in intervals. If an I/O passes the limits of such an interval, an event is generated. It is assumed that continuous state signals are available and can also be quantized. The idea is to identify a timed automaton based on the quantized I/O and state signals for the normal and for some faulty behavior. The quantized state signals determine the state space of the timed automaton. To get the necessary samples of the faulty system behavior it must be possible to introduce faults and to observe the resulting behavior. For each fault, an own timed automaton is identified. The set of identified automata is used for online fault diagnosis by analyzing which automaton can reproduce the currently observed system behavior. If the observation can only be reproduced by an automaton identified on the basis of some faulty system behavior, a fault is detected.

The class of automata used in (Supavatanakul et al., 2006) is similar to Mealy automata with inputs and outputs associated to transitions. Additionally, the *timed* behavior is defined by time intervals added to the transitions. A transition can only fire if the automaton clock is within this time interval. If x_1 is the current state of the automaton in figure 2.16 and the current clock value is 5, it is possible that the transition from x_1 to x_2 fires reading the input $a = 1$ and producing the output $b = 2$ since the current value of the automaton clock is within the given time interval. The transition to x_3 cannot fire since the clock already exceeded its time interval.

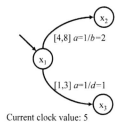

Figure 2.16: Example for a timed automaton

The principle of the (Supavatanakul et al., 2006) identification approach can be seen in figure 2.17. In the example, two input signals a and c and two output signals b and d exist. In this simplified example, the state signals are equal to the input and output signals. Three timed sequences can be seen that are used to identify the automaton in the upper part of figure 2.17. Each sequence starts with the same setting of I/O values defining state x_1. In the first sequence, $a = 1/b = 1$ occurs after four time units leading to the I/O value setting defining state x_2. Hence, x_1 and x_2 are connected. The time interval at the transition between x_1 and x_2 after sequence 1 is $[4, 4]$. The rest of the first sequence is processed following the same principle: if an I/O value setting occurs that is not yet represented by an own state, a new state is created and connected to the state representing the former setting. Time intervals of newly created states contain the number of elapsed time units since the last I/O change as single value. After the analysis of sequence 1, the automaton consists of the state trajectory x_1, x_2, x_3 and x_4. Processing the second sequence does not lead to an enlarged state space, since the occurring I/O value settings are already part of the automaton. Since the *timed* behavior of this trajectory differs from the first one, the time intervals of the automaton are adapted. $a = 1/b = 1$ for example occurs after eight time units. Hence, the interval of the transition from x_1 to x_2 is redefined to $[4, 8]$. The analysis of the remaining part of sequence 2 and of sequence 3 leads to the automaton shown in figure 2.17.

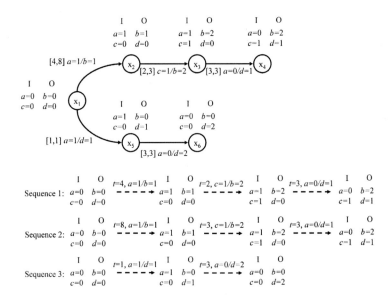

Figure 2.17: Example for the (Supavatanakul et al., 2006) approach

Although the approach of (Supavatanakul et al., 2006) has been successfully applied to a benchmark system, its use poses some problems. The first problem is that it is necessary to identify fault-free and faulty system behavior. In large closed-loop DES it is not possible to artificially introduce each possible fault at each possible time to deliver the necessary observed system data. Hence, it is difficult to guarantee that the identified models contain the whole possible faulty and fault-free system behavior. The second problem of the approach is related to the determination of time interval limits. The limits are determined by the shortest and the longest state duration. If a maximum time limit is fixed to 10 time units, a behavior occurring after 11 time units will not be recognized. It is difficult to decide if the shortest and longest durations have already been observed or if more extreme values are possible for the considered faulty or non-faulty scenarios.

2.4.3 Identification of Petri nets

Apart from identifying finite automata to represent a DES language, various approaches to determine Petri nets based on observed event sequences exist. In (Pia Fanti and Seatzu, 2008) an overview of the most important ones is given. In this section, two representative approaches for the main ideas of Petri net identification are explained. The first idea is to use rules for a systematic construction of a Petri net from observed sequences. The second idea is to solve an integer linear programming problem over structural constraints such that the identified Petri net can reproduce the observed event sequences.

An approach which is representative for the first class of methods is given in (Meda-Campana and Lopez-Mellado, 2005). The approach works with interpreted Petri nets (IPN, see section 2.3.2). An IPN can be seen as a labeled Petri net where the events associated to the transitions represent the system input and a special function φ determines the output based on the marking of the IPN. The IPN has unobservable and observable places. The data base for the approach of (Meda-Campana and Lopez-Mellado, 2005) are sequences of observed transitions and place markings. The idea is to build so-called cyclic m-words beginning and ending with the same marking $\varphi(\mathbf{X})$ of observable places. For each pair of transitions in two observed m-words like $w_1 = t_1, t_2, t_3, t_4, t_5$ and $w_2 = t_3, t_1, t_2, t_4, t_5$ it is determined if they are *dependent* or *concurrent*. Two transitions t_i, t_j are dependent if there exists a place p_k such that there is an arc leading from t_i to p_k and if there is an arc leading from p_k to t_j. If two transitions always occur in the same order, they are possibly dependent (like t_4 and t_5 in w_1 and w_2). Two transitions t_i, t_j are concurrent if they can occur in different orders like t_2 and t_3 in w_1 and w_2. Figure 2.18 shows an example for the identification approach of (Meda-Campana and Lopez-Mellado, 2005). The shaded Petri net places are supposed to be unobservable. The leftmost Petri net is the (unknown) original DES to be identified. The Petri net in the middle is constructed upon evaluation of the m-word $w_1 = t_1, t_2, t_3, t_4, t_5$. In the example, the resulting system markings $\varphi(\mathbf{X})$ are not shown, but can be derived from the original Petri net in the left of figure 2.18. After an analysis of w_1, various pairs of transitions are supposed to be dependent (e.g. t_1, t_2 or t_2, t_3). Since two dependent transitions must always be connected by a place, the resulting Petri net is a simple chain of places and transitions. The resulting marking $\varphi(\mathbf{X})$ between t_2 and t_3 leads to the conclusion that no observable place contains a token. As a consequence, the unobservable place p_6 is added. Place p_5 from the original Petri net is not added to the identified model since t_2 and t_5 are no direct successors. Hence, there is no dependency which must be translated by a place between the two transitions.

After the observation of $w_2 = t_3, t_1, t_2, t_4, t_5$, the Petri net structure is altered. Comparing w_1 and w_2, it can be concluded that t_1, t_3 and t_2, t_3 are concurrent since they occur in different orders. The concurrency of t_2, t_3 shows that the two transitions are not dependent like assumed after analyzing w_1. Hence, place p_6 can be removed such that t_2 and t_3 can fire concurrently. In (Meda-Campana and Lopez-Mellado, 2005) it is shown, that if a sufficiently large and divers set of words has been observed, the identified model becomes more and more similar to the original system.

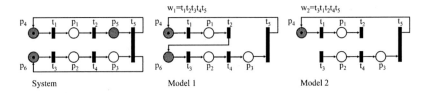

Figure 2.18: Example for the (Meda-Campana and Lopez-Mellado, 2005) approach

The approach of (Meda-Campana and Lopez-Mellado, 2005) poses two problems. The first one is its inability to guarantee the accuracy of the identified model. Model 2 in figure 2.18 e.g. can also generate the word $w_3 = t_3, t_4, t_1, t_5, t_2$ which is not part of the original system language specified by the left most Petri net. The second problem arises from the concurrency analysis. Each pair of transitions must be kept in memory to analyze if it already occurred in different orders. If a system with many transitions and a huge number of long m-words is to be identified, the concurrency check can significantly decrease the performance of the identification algorithm.

A different identification approach is proposed in (Dotoli et al., 2008). The approach works with so-called λ-free labeled Petri nets which have a different semantic as the Petri nets defined in section 2.2.3. The difference is the explicit definition of the pre- and post-incidence matrices **Pre** and **Post**:

Definition 13 (Petri net according to (Dotoli et al., 2008)). A Petri net is a bipartite graph described by the four-tuple $PN = (P, T, \textbf{Pre}, \textbf{Post})$ where P is a set of places with cardinality m, T is a set of transitions with cardinality n, $\textbf{Pre} : P \times T \to \mathbb{N}^{m \times n}$ and $\textbf{Post} : P \times T \to \mathbb{N}^{m \times n}$ are the pre- and post-incidence matrices, respectively, which specify the arcs connecting places and transitions. More precisely, for each $p \in P$ and $t \in T$ element $\textbf{Post}(p,t)$ $(\textbf{Pre}(p,t))$ is equal to a natural number indicating the arc multiplicity if an arc going from p to t (from t to p) exists, and it equals 0 otherwise.

The state of a Petri net is determined by its current marking $\textbf{M} : P \to \mathbb{N}^m$ assigning to each place of the net a nonnegative number of tokens. A Petri net place can be observable or unobservable. Observable places are part of the output vector $\textbf{y}$. A Petri net system $< PN, \textbf{M}_0 >$ is a net PN with an initial marking $\textbf{M}_0$. The incidence matrix of the Petri net is defined as $\textbf{C} = \textbf{Post} - \textbf{Pre}$. A transition $t_j \in T$ is enabled at a marking $\textbf{M}$ if and only if for each pre-place p of t_j $(\textbf{Pre}(p, t_j) > 0)$, $\textbf{M}(p) \geq \textbf{Pre}(p, t_j)$ holds. The incoming places of t_j must have enough tokens to enable a firing of the according transition. If transition t_j fires, it produces a new marking given by $\textbf{M}' = \textbf{M} + \textbf{C}\vec{t}_j$ where $\vec{t}_j$ is the n-dimensional firing vector corresponding to the j-th canonical basis vector. A sequence of firing transitions is defined as $\sigma = t_{\beta_1}, t_{\beta_2}, \ldots, t_{\beta_h}$. The Petri net system can generate a language using a labeling function $\lambda : T \to E$ which assigns to each transition $t \in T$ a symbol $e_i \in E$. The labeling function is called λ-free if the same label $e_i \in E$ may be associated to more than one transition while no transition may be labeled with the empty string ε. $\lambda(\sigma)$ denotes the event sequences produced during the firing sequence σ. The language $\mathcal{L}(PN, \textbf{M}_0)$ is defined by the output of sequences of transition firings starting in $\mathbf{M}_0$.

The data basis of the identification approach of (Dotoli et al., 2008) is an observed sequence $w = e_{\alpha_1}e_{\alpha_2}\ldots e_{\alpha_h} \in \mathcal{L}$ and the corresponding output vectors $\mathbf{y}$. Before the identification algorithm starts, an upper bound m for the cardinality of the place set P and an upper bound n for the the transitions set T must be given. Additionally, the labeling function λ has to be predefined. With this information it is possible to determine the firing sequence $\sigma = t_{\beta_1}^{\alpha_1}t_{\beta_2}^{\alpha_2},\ldots,t_{\beta_h}^{\alpha_h}$ with $\lambda(\sigma) = w$. The idea of (Dotoli et al., 2008) is to identify a Petri net using integer linear programming. It is shown that each Petri net being capable of producing the observed data with at most m places and n transitions must satisfy the following constraint:

$$\xi(w,\mathbf{y},\lambda,T,m) = \begin{cases} \mathbf{Pre}, \mathbf{Post} \in \mathbb{N}^{m \times n} \\ \mathbf{M}_i \in \mathbb{N}^m \text{ with } i = 0,\ldots,h \\ \mathbf{Post}^T\vec{\mathbf{1}}_{m \times 1} + \mathbf{Pre}^T\vec{\mathbf{1}}_{m \times 1} \geq \vec{\mathbf{1}}_{n \times 1} \\ \mathbf{Post} \cdot \vec{\mathbf{1}}_{n \times 1} + \mathbf{Pre} \cdot \vec{\mathbf{1}}_{n \times 1} \geq \vec{\mathbf{1}}_{m \times 1} \\ \forall t_{\beta_i}^{\alpha_i} \in \sigma \text{ with } \lambda(\sigma) = w : \mathbf{Pre} \cdot \vec{\mathbf{t}}_{\beta_i}^{\alpha_i} \leq \mathbf{M}_{i-1} \\ \forall t_{\beta_i}^{\alpha_i} \in \sigma \text{ with } \lambda(\sigma) = w : (\mathbf{Post} - \mathbf{Pre}) \cdot \vec{\mathbf{t}}_{\beta_i}^{\alpha_i} = \mathbf{M}_i - \mathbf{M}_{i-1} \end{cases}$$

where $\vec{\mathbf{1}}_{m \times n}$ is the matrix of dimensions $m \times n$ with each element being 1. The first two elements of the constraint assure that the resulting PN system has at most m places and n transitions. The constraint $\mathbf{Post}^T\vec{\mathbf{1}}_{m \times 1} + \mathbf{Pre}^T\vec{\mathbf{1}}_{m \times 1} \geq \vec{\mathbf{1}}_{n \times 1}$ makes sure that the Petri net does not have transitions that are not connected to any place. Analogously, $\mathbf{Post} \cdot \vec{\mathbf{1}}_{n \times 1} + \mathbf{Pre} \cdot \vec{\mathbf{1}}_{n \times 1} \geq \vec{\mathbf{1}}_{m \times 1}$ assures that the Petri net does not have any places which are not connected to at least one transition. The fifth constraint makes sure that each transition in the firing sequence is enabled by its preceding marking. The last constraint represents the fact that two successive markings must be calculated by the incidence matrix and the corresponding transitions $\mathbf{M}_i = \mathbf{M}_{i-1} + (\mathbf{Post} - \mathbf{Pre}) \cdot \vec{\mathbf{t}}_{\beta_i}^{\alpha_i}$.

This constraint represents the necessary condition of a Petri net which is able to reproduce the observed data. Generally, there are many Petri nets for which $\xi(w,\mathbf{y},\lambda,T,m)$ holds. To select one solution, it is possible to use a metric which can be minimized by the integer linear programming problem solver. In (Dotoli et al., 2008), the following function is given as an example:

$$\phi(\mathbf{Pre}, \mathbf{Post}, \mathbf{M}_0) = \vec{\mathbf{a}}^T \cdot \mathbf{Pre} \cdot \vec{\mathbf{b}} + \vec{\mathbf{c}}^T \cdot \mathbf{Post} \cdot \vec{\mathbf{d}} + \vec{\mathbf{e}}^T \cdot \mathbf{M}_0$$

where the vectors $\vec{\mathbf{a}}, \vec{\mathbf{b}}, \vec{\mathbf{e}} \in \mathbb{N}^m$ and $\vec{\mathbf{b}}, \vec{\mathbf{d}} \in \mathbb{N}^n$ are free parameters to adjust the metric to a given purpose. One possibility is for example to chose the parameters such that the resulting Petri net has a minimal number of initial tokens and arcs:

$$\phi(\mathbf{Pre}, \mathbf{Post}, \mathbf{M}_0) = \vec{\mathbf{1}}_{1 \times m} \cdot (\mathbf{Pre} + \mathbf{Post}) \cdot \vec{\mathbf{1}}_{n \times 1} + \vec{\mathbf{1}}_{1 \times m} \cdot \mathbf{M}_0$$

In this case, both optimization objectives have the same weight. After formulation of the optimization objectives and constraints, the identification problem can be solved with an integer linear programming problem solver (ILP solver).

The advantage of the optimization approach is that it guarantees the completeness of the identified models: the constraint assures that each observed sequence is part of the identified language. A disadvantage is that for large systems an ILP solver takes long to admit a solution. The second main disadvantage is that the formulation of the objective function is not trivial. If several objectives have to be met (e.g. model size and accuracy), it is a difficult problem to find appropriate parameters for the objective function.

Nevertheless, other approaches with the same main idea like (Giua and Seatzu, 2005) show that using optimization techniques to solve the identification problem for DES is a promising approach.

2.4.4 Discussion

The general considerations in section 2.4.1 showed that special accuracy demands for identified models exist if they are to be used for online fault detection. The first constraint is that the observed part of the original system language must be reproducible by the identified model. It was shown that this minimizes the number of false alerts during online fault diagnosis with an identified model. Except of the optimization-based automaton identification from (Baron et al., 2001b), all presented methods deliver models which are able to completely reproduce the observed system data. The second main constraint is that the identified language should not contain an important part of sequences not being part of the original system language. This special constraint is not met by any of the presented identification methods. It is not possible to restrict the identified language in a well defined way such that the necessary accuracy can be guaranteed.

For automata and for Petri nets two main identification ideas can be distinguished. The first idea is to use rules coded in an algorithm which processes the observed system data. These approaches (like (Supavatanakul et al., 2006) for automata or (Meda-Campana and Lopez-Mellado, 2005) for Petri nets) deliver a model in a deterministic way since the available system data is always treated in the same manner. The second major approach for identification of DES is to use optimization techniques like (Baron et al., 2001b) for automata or (Dotoli et al., 2008) for Petri nets. The identification objective must be formalized in an objective function which is given to an optimization algorithm. A challenge in optimization-based identification is the choice of an appropriate fitness function which needs a considerable expertise of the optimization method and of the mathematics describing the identification problem. The outcome of the optimization approaches can vary if heuristic optimization techniques like genetic algorithms or heuristic solvers for integer linear programming problems are used. Another important drawback of identification methods based on optimization techniques is that they are expensive in calculation time. Although it has been shown that they work with academic examples or small case studies (Dotoli et al., 2006), they have not been applied to real-world systems.

Since non of the presented methods guarantees the necessary accuracy for model-based fault diagnosis in closed-loop DES, an identification method especially suited for this domain has been proposed in (Klein, 2005). The method delivers a model in form of a monolithic automaton which guarantees some important accuracy and completeness properties. Under some conditions, the method delivers an appropriate model for online fault diagnosis. The model is determined by a deterministic identification algorithm which will be introduced in the next chapter. It is the basis for the presented work. In chapter 5 the idea of using optimization approaches for identification in DES will be embraced to improve the algorithm from (Klein, 2005). The resulting distributed identification will thus combine the idea of using deterministic identification algorithms and the idea of using optimization techniques.

3 Monolithic Identification of Closed-Loop Discrete Event Systems

3.1 Model class

In the previous chapter it has been explained that none of the existing identification methods delivers a model which is appropriate for online fault diagnosis in industrial closed-loop DES. One of the main reasons is that the identified models do not appropriately represent the special characteristics of this class of systems. As explained in section 2.1, a closed-loop DES consists of a plant and a controller as shown in figure 3.1. In the remaining part of this work, the notion closed-loop DES always refers to the fault-free system if not stated otherwise. The system characteristics have an important influence on the choice of an appropriate model class and are discussed before an identification algorithm is given.

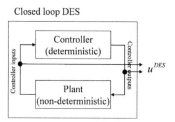

Figure 3.1: Structure of the considered systems

Observable events produced in a closed-loop DES are displayed in the data exchanged between controller and plant. This data is considered as the system output in this work. In figure 3.1, the output is represented by the symbol u^{DES}. The original system language L_{Orig} (see section 2.4.1) consists of sequences of output symbols u^{DES}. Since a closed-loop system does not have any inputs but exhibits an event-driven series of outputs, it can be interpreted as a non-deterministic autonomous event generator. The controller of the closed-loop DES is usually designed as a deterministic subsystem: if the controller is in a given state and a controller input sequence is applied several times from that state, the result are identical controller output sequences. For the plant, this property of determinism does not hold. Applying the same sequence of actuator commands several times to the plant being in a given state does not necessarily reproduce exactly the same sequence of delivered sensor signals. The varying sequences of sensor signals are the result of physical components like motors, cylinders or valves that do not always react exactly in the same way. If two or more components are triggered simultaneously, the order of the reaction may change from one triggering to another depending on external influences like friction or mechanical constraints. These influences are not predictable and must be considered as non-determinisms. Coupling a deterministic and a non deterministic system as in figure 3.1 leads to a non-deterministic closed loop system.

If no information about the controller algorithm or the plant structure is used, a closed loop DES can be seen as a non-deterministic autonomous black-box. An appropriate way to represent the observable behavior of such systems are finite state machines that reflect the characteristics of a black-box event generator. This necessitates the introduction of the notion of 'state' for the description of the system behavior. In figure 3.2, a possible finite state machine representation for the considered closed-loop DES is shown. The figure contains a non-deterministic autonomous event generator determining the future evolution and the external behavior. The consequences of an event are given in the following definition:

Definition 14 (Event produced by a closed-loop DES). An event produced by a closed-loop DES leads to a new current system state x^{DES}. With the notation $x^{DES}(j)$, the current state after the occurrence of the j-th event is denoted.

The current system state produces an output u^{DES} via the function Λ and is determined by the last system state without considering any input data by the function F_N. In case of closed-loop DES, the system state x^{DES} represents the combined state of controller and plant.

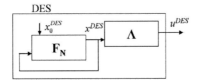

Figure 3.2: Autonomous state-based event generator

Formally, the current system state is determined using a non-deterministic function $F_N : X^{DES} \rightarrow 2^{X^{DES}}$ with $2^{X^{DES}}$ representing the powerset[1] of the DES state space X^{DES}.

Definition 15 (External behavior of closed-loop DES). The external behavior of a closed-loop DES is described by its current state $x^{DES}(j) \in F_N(x^{DES}(j-1))$ and its output $u^{DES}(j) = \Lambda(x^{DES}(j))$ which can be observed after the j-th event. If there are several next state candidates $\left(|F_N(x^{DES}(j-1))| > 1\right)^{2}$ the current state $x^{DES}(j)$ is chosen non-deterministically from the set $F_N(x^{DES}(j-1))$ upon the occurrence of the j-th event. x_0^{DES} denotes the initial system state.

---

[1]Example for the powerset of $A = \{a, b, c\} : 2^A = \{\{\}, \{a\}, \{b\}, \{c\}, \{a, b\}, \{a, c\}, \{b, c\}, \{a, b, c\}\}$
[2]With $|Set|$, the set cardinality is denoted

From this definition it follows directly that each event is translated into a new system output u^{DES}. It is assumed that a new system output differs from its predecessor (only observable events are considered):

Assumption 1 (Distinguishable system outputs). It is assumed that the j-th event leads to a state with a new output such that $u^{DES}(j) \neq u^{DES}(j-1)$.

In (Klein, 2005) a model which is able to represent a system like in figure 3.2 is defined as a **N**on-**D**eterministic **A**utonomous **A**utomaton with **O**utput (NDAAO).

Definition 16 (Non-deterministic autonomous automaton with output (NDAAO)). $NDAAO = (X, \Omega, f, \lambda, x_0)$ with X finite set of states, Ω output alphabet, $f : X \to 2^X$ non-deterministic transition function, $\lambda : X \to \Omega$ output function and x_0 the initial state[3].

An NDAAO is able to describe a closed-loop DES like in figure 3.2 if it can produce the same external behavior and if it follows the same dynamics. It is obvious that the first condition holds if the current NDAAO state is determined using the condition $x(j) \in f(x(j-1))$ which corresponds to $x^{DES}(j) \in F_N(x^{DES}(j-1))$ and if the current output $u(j)$ is calculated with $u(j) = \lambda(x(j))$ which corresponds to $u^{DES}(j) = \Lambda(x^{DES}(j))$. The NDAAO dynamics are similar to the closed-loop DES dynamics if they are defined in accordance with definition 15:

Definition 17 (Next-state rule NDAAO). An NDAAO state $x(j)$ can be the successor of the last current state $x(j-1)$ if and only if $x(j) \in f(x(j-1))$. If there are several next state candidates ($|f(x(j-1))| > 1$) the new actual state $x(j)$ is chosen non-deterministically from the set $f(x(j-1))$.

To formulate an identification algorithm which delivers an NDAAO for a given closed-loop DES it is necessary to define the data basis which is used for identification. This data base is given by the language[4] of the closed-loop DES observed during a certain amount of *system evolutions*. In manufacturing systems such an evolution can be a production cycle that is carried out by the system.

When the closed-loop DES performs an evolution, sequences of outputs u can be observed. The identification is carried out on the basis of p different system evolutions. Since each change in value of the output u is considered as the result of an event, the output after the j-th event is given by the next definition:

Definition 18 (System output). The j-th output in the h-th of p system evolutions is defined as $u_h(j)$.

The sequence that is observed when an evolution is performed is built by the outputs u in the order of their occurrence.

---

[3]The NDAAO-definition in (Klein, 2005) also contains a final state. After some modifications in the identification algorithm (see section 3.2), this state is no longer necessary.

[4]The language definitions in this section are modified versions of the definitions in (Klein, 2005) but follow the same ideas.

Definition 19 (Observed sequence). If during the h-th system evolution l_h outputs u_h have been observed, the sequence is denoted as $\sigma_h = (u_h(1), u_h(2), \ldots, u_h(l_h))$. The set of all observations is denoted as $\Sigma = \{\sigma_1, \ldots, \sigma_p\}$.

Assumption 2 (Initial system output). It is assumed that each observed sequence starts with the same output symbol: $\sigma_i(1) = \sigma_j(1)$ holds $\forall \sigma_i, \sigma_j \in \Sigma$.

Based on the observed sequences it is possible to define the language of the considered systems.

Definition 20 (Observed word set and language). The words of length q observed during p different system evolutions are denoted as

$$W_{Obs}^q = \bigcup_{\sigma_h \in \Sigma} \left(\bigcup_{j=1}^{|\sigma_h|-q+1} (u_h(j), u_h(j+1), \ldots, u_h(j+q-1)) \right).$$

$|\sigma_h|$ denotes the length of the h-th observed sequence. With the observed word set the observed system language of length n is defined as

$$L_{Obs}^n = \bigcup_{i=1}^{n} W_{Obs}^i$$

The language thus consists of sequences of output symbols.

Remark 1. From definition 20 and assumption 1 it follows directly that for two successive $(u_h(j), u_h(j+1)) \in L_{Obs}^n$ it holds: $u_h(j) \neq u_h(j+1)$.

Assumption 3 (Asymptotic convergence of L_{Obs}^n to L_{Orig}^n). It is assumed that the observed language L_{Obs}^n is exhibited by the system such that it (asymptotically) converges to the complete original system language L_{Orig}^n (which is assumed to be finite) when increasing the number of observed evolutions h.

Despite the autonomous nature of the considered class of systems we assume that with a growing observation time the original language is more and more completely exhibited and can thus be observed. Since different system evolutions are supposed to be similar (not necessarily equal), the observed language is supposed to converge to the original system language with rising observation time like sketched in figure 3.3 although it can take many evolutions to reach convergence. When the case study system and the industrial application are treated, it will be shown that the assumed evolution of the observed language cardinality is realistic for existing closed-loop DES.

The aim of identification is to determine a model which is able to approximate the system language. Hence, the language of the NDAAO must also be defined.

Definition 21 (Word set and language of the NDAAO). The set of words of length n generated from a state $x(i)$ is defined as:

$$W_{x(i)}^{n=1} = \{w \in \Omega^1 \mid w = \lambda(x(i))\}$$

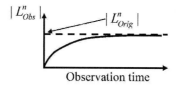

Figure 3.3: Assumed evolution of the observed language cardinality

and

$$W_{x(i)}^{n>1} = \{w \in \Omega^n \mid \big(w = (\lambda(x(i)), \lambda(x(i+1)), \ldots, \lambda(x(i+n-1)))\big) \wedge$$
$$x(j+1) \in f(x(j)) \forall i \leq j < i+n-1)\}$$

The language generated by the NDAAO is given by

$$L_{Ident}^n = \bigcup_{i=1}^{n} \bigcup_{x \in X} W_x^i$$

If the output alphabet Ω consists of the observed outputs u^{DES}, the NDAAO can reproduce the observed system language by performing *state trajectories*. With definition 20 and 21, the observed language L_{Obs}^n and the identified language L_{Ident}^n are prefix-closed (Cassandras and Lafortune, 2006).

The purpose of the identified NDAAO is to be used in online fault-diagnosis. It will be used as a model of the *fault-free* system behavior. This motivates the following assumption:

Assumption 4 (Fault-free observation). For the use of an identified NDAAO in online fault diagnosis, it is assumed that the observed system language L_{Obs}^n is a subset of the *fault-free* original system language L_{Orig}^n.

3.2 Identification algorithm

Based on the definitions of the former section, it is possible to give an identification algorithm that builds an NDAAO by an analysis of the observed language. The original version of the following algorithm has been given in (Klein, 2005). In this section, a modified version is introduced. The main difference is that the reformulated algorithm is no longer restricted to system evolutions starting and ending with the same system output.

In order to adjust the model accuracy with respect to the observed system language, the identification algorithm has a free parameter k. It defines the length of words which is used as identification data base. The higher k is chosen, the more reliably the identified model can reproduce the observed system language. In section 3.3 it will be shown that the parameter helps to guarantee important accuracy and completeness properties of the identified model. The question how to chose an appropriate value for k will be discussed in section 3.4.

The first step of the identification procedure is to modify the observed sequences. The first output symbol in each sequence is duplicated $k-1$ times:

$$\sigma_h^k(i) = \begin{cases} \sigma_h(1) \text{ for } 1 \leq i \leq k \\ \sigma_h(i-k+1) \text{ for } k < i \leq k + |\sigma_h| - 1 \end{cases} \tag{3.1}$$

$|\sigma_h|$ denotes the length of the sequence σ_h. Equation 3.1 is applied to each sequence $\sigma_h \in \Sigma$ resulting in $\Sigma^k = \{\sigma_1^k, \ldots, \sigma_p^k\}$. On the basis of Σ^k, the modified word sets of length k and $k+1$ are determined according to definition 20:

$$W_{Obs,\Sigma^k}^k = \bigcup_{\sigma_h^k \in \Sigma^k} \left(\bigcup_{i=1}^{|\sigma_h^k|-k+1} (u_h(i), u_h(i+1), \ldots, u_h(i+k-1)) \right) \tag{3.2}$$

$$W_{Obs,\Sigma^k}^{k+1} = \bigcup_{\sigma_h^k \in \Sigma^k} \left(\bigcup_{i=1}^{|\sigma_h^k|-k} (u_h(i), u_h(i+1), \ldots, u_h(i+k)) \right) \tag{3.3}$$

W_{Obs,Σ^k}^k and W_{Obs,Σ^k}^{k+1} are the data basis for algorithm 1 which identifies an NDAAO. For the algorithm, we define an operator $w^q \langle a..b \rangle$ to deliver the substring from position a to position b in word w^q:

$$w^q \langle a..b \rangle := (u(a), u(a+t), \ldots, u(b)) \forall 1 \leq t < b - a \tag{3.4}$$

with $w^q \in W_{Obs}^q : w^q = (u(1), \ldots, u(q))$ and $a \geq 1, b \leq q$

As a special case, we define that with $w^q \langle a \rangle$ it is possible to determine the symbol at the position a in w^q.

If $w^6 = ABCDEF$ and $a = 2, b = 4$, $w^6 \langle a..b \rangle = BCD$.

In the first step of the identification algorithm, the state space of the NDAAO is created. For each word $w^k \in W_{Obs,\Sigma^k}^k$ a state x is built and its output function is associated with this word ($\lambda(x) := w^k$). After this step, the automaton consists of isolated, non-connected states. The states get connected in step 2. In this step, the transition function is built on the basis of W_{Obs,Σ^k}^{k+1}. For each word w^{k+1} in this set the first k-long subsequence $w^{k+1} \langle 1..k \rangle$ and the last k-long subsequence $w^{k+1} \langle 2..k+1 \rangle$ are determined with the operator from equation 3.4. From step 1, for each substring there exists exactly one state which has this string as output. The two according states x and x' are selected and get connected such that the state representing the first substring gets the state of the second substring added to its successor states ($f(x) := f(x) \cup x'$). Due to step 1, each state has a word of length k as output. In step 3, the state representing the sequence with the $k-1$-times duplicated first output symbols becomes the initial state. By assumption 2, each observed sequence starts with the same output. With step 1, there is only one state representing the $k-1$-times duplicated first output symbols. For the construction of the automaton language according to definition 21 such that it reproduces the language of the closed-loop DES, it is only necessary to have a single output symbol as state output. Hence, in step 4 the output function of the states is reassigned with the last 'letter' of the output word $(\lambda(x) := \lambda(x)\langle|\lambda(x)|\rangle)$[5]. To reduce the state space, in the last step of the identification algorithm, equivalent states are merged. According to (Cassandras and Lafortune, 2006), two automaton states x_1, x_2 are equivalent if $\forall n : L_{x_1}^n = L_{x_2}^n$. In (Klein, 2005) it is shown that this condition is fulfilled for two NDAAO-states if they are associated with the same output, i.e. $\lambda(x_1) = \lambda(x_2)$ and if they have the same set of following states, i.e. $f(x_1) = f(x_2)$. A procedure to merge equivalent states is given in algorithm 2.

Algorithm 1 Monolithic identification algorithm

Require: Parameter k, observed word sets W_{Obs,Σ^k}^k and W_{Obs,Σ^k}^{k+1}

1: $X := \{(\forall w^k \in W_{Obs,\Sigma^k}^k)(\exists! x) \wedge (\lambda(x) := w^k, f(x) := \{\})\}$

2: $(\forall (x, x', w^{k+1}) \in X \times X \times W_{Obs,\Sigma^k}^{k+1})|(\lambda(x) = w^{k+1}\langle 1..k\rangle \wedge \lambda(x') = w^{k+1}\langle 2..k+1\rangle)$ do
   $f(x) := f(x) \cup x'$

3: $x_0 := x \in X|(\lambda(x) = w^k$ and $w^k\langle i\rangle = \sigma_1(1)(\forall 1 \leq i \leq k))$

4: $(\forall x \in X) : \lambda(x) := \lambda(x)\langle|\lambda(x)|\rangle$

5: Merge equivalent states in X with algorithm 2

Algorithm 2 performs for each state x_i in the NDAAO state space the following operations: First, a set of equivalent states is built in line 2. Each state in $X_{x_i,Eq}$ has the same output and the same following states like the considered state x_i. The equivalent states will be removed from the state set. Hence, in the loop in line 3 and 4, each state x_{pre} that has one of the equivalent states as following state is updated such that its transition function no longer contains one of the states that will be removed. Instead, the currently considered state x_i is added to their transition function. In line 6 it is checked if the set of equivalent states was empty (the currently considered state x_i does not have equivalent states). If yes, the next state from the state set is considered in line 1. If $X_{x_i,Eq}$ is not empty, the states in this set are removed from X in line 7 and the algorithm starts again. It now considers the updated state set that does no longer contain the equivalent states of x_i from the former run of the algorithm. In (Klein, 2005) is is shown that merging of equivalent states does not influence the language created by the identified NDAAO.

The principles of the algorithms are now shown with an example. It is assumed that the following three sequences have been observed: $\sigma_1 = (A, B, C, D, E, A)$, $\sigma_2 = (A, D, B, C, D, A, C, A)$ and $\sigma_3 = (A, D, B, C, F, D, E, B)$. For the example, the identification parameter k is chosen to $k = 2$. First, equation 3.1 is applied to the three sequences to duplicate the first output symbol $k - 1$ times. The result is

$$\sigma_1^{k=2} = (A, A, B, C, D, E, A)$$
$$\sigma_2^{k=2} = (A, A, D, B, C, D, A, C, A)$$
$$\sigma_3^{k=2} = (A, A, D, B, C, F, D, E, B)$$

---

[5]$|w|$ denotes the length of word w: $|ABC| = 3$. Since $\lambda(x)$ is a word of length k until this step, $\lambda(x) := \lambda(x)\langle|\lambda(x)|\rangle$ selects the last letter of this word and takes it as new state output.

Algorithm 2 State merging algorithm

Require: State space X

1: **for all** $x_i \in X$ **do**

2: $\quad X_{x_i,Eq} := \{(x' \in X \backslash x_i)| \ (\lambda(x_i) = \lambda(x') \wedge f(x_i) = f(x'))\}$

3: $\quad$ **for all** $(x_{pre} \in X)|(\exists x_{eq} \in X_{x_i,Eq}) \wedge (x_{eq} \in f(x_{pre}))$ **do**

4: $\quad\quad f(x_{pre}) := (f(x_{pre}) \backslash x_{eq}) \cup x_i$

5: $\quad$ **end for**

6: $\quad$ **if** $X_{x_i,Eq} \neq \{\}$ **then**

7: $\quad\quad X := X \backslash X_{x_i,Eq}$

8: $\quad\quad$ Restart Algorithm with new X

9: $\quad$ **end if**

10: **end for**

The next step is to determine W^k_{Obs,Σ^k} and W^{k+1}_{Obs,Σ^k} according to equations 3.2 and 3.3. The results are

$$W^2_{Obs,\Sigma^2} = \{AA, AB, BC, CD, DE, EA, EB, AD, DB, DA, AC, CA, CF, FD\}$$

and

$$W^3_{Obs,\Sigma^2} = \{AAB, ABC, BCD, CDE, DEA, DEB, AAD, ADB,$$
$$DBC, CDA, DAC, ACA, BCF, CFD, FDE\}$$

At this point, the necessary data for algorithm 1 is available. Figure 3.4 shows the result of step 1 and 2. For each word in W^2_{Obs,Σ^2}, a state has been created. Based on W^3_{Obs,Σ^2}, the states have been connected. State x_0 for example has been connected to states x_1 and x_6 because of the words $w^3 = AAB$ and $w^3 = AAD$. $w^3 = AAB$ was divided in the two substrings $w^3\langle 1..k \rangle = w^3\langle 1..2 \rangle = AA$ and $w^3\langle 2..k+1 \rangle = w^3\langle 2..3 \rangle = AB$. Hence, states x_0 with $\lambda(x_0) = AA$ and x_1 with $\lambda(x_1) = AB$ have been connected. In step 4 of algorithm 2 the output function is redefined. The new output function is the last letter of former state output. In figure 3.4, the state outputs are replaced by the bold/italic letters.

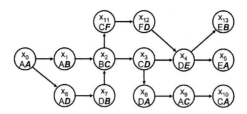

Figure 3.4: Identified NDAAO after step 2 and 3

The resulting automaton is then given to the state merging algorithm. It is determined that states x_1 and x_7 as well as x_5 and x_{10} have the same output and following

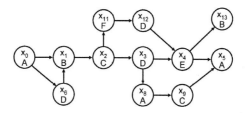

Figure 3.5: The identified NDAAO after merging of equivalent states

states. Hence, they are equivalent and can be merged. The resulting automaton is given in figure 3.5.

The identified automaton in figure 3.5 can produce the following ten sequences starting in x_0 and ending in x_5 or x_{13}:

$$\sigma_1^{NDAAO} = (A, B, C, D, E, A) \qquad \sigma_2^{NDAAO} = (A, B, C, F, D, E, A)$$
$$\sigma_3^{NDAAO} = (A, B, C, D, A, C, A) \qquad \sigma_4^{NDAAO} = (A, D, B, C, F, D, E, A)$$
$$\sigma_5^{NDAAO} = (A, D, B, C, D, E, A) \qquad \sigma_6^{NDAAO} = (A, D, B, C, D, A, C, A)$$
$$\sigma_7^{NDAAO} = (A, B, C, D, E, B) \qquad \sigma_8^{NDAAO} = (A, D, B, C, D, E, B)$$
$$\sigma_9^{NDAAO} = (A, B, C, F, D, E, B) \qquad \sigma_{10}^{NDAAO} = (A, D, B, C, F, D, E, B)$$

Three of these sequences correspond to observed ones: $\sigma_1^{NDAAO} = \sigma_1$, $\sigma_6^{NDAAO} = \sigma_2$ and $\sigma_{10}^{NDAAO} = \sigma_3$.

3.3 Important properties of the identified automaton

An NDAAO identified with algorithm 1 has some important properties which make the automaton appropriate for online fault diagnosis. The first important property is an accuracy guarantee which has already been shown in (Klein, 2005): For a given value of the identification parameter k, the identified NDAAO is $(k + 1)$-complete (Moor et al., 1998), i.e. $\forall n \leq k + 1$, $L_{Ident}^n = L_{Obs}^n$. This guarantees that the automaton minimizes the exceeding language of length $k + 1$ and thus reduces the number of non-detectable faults (see section 2.4.1). This property will be shown in theorem 4 on the following pages. Since the algorithm of (Klein, 2005) has been modified, it is necessary to show that the reformulated algorithm also guarantees the model accuracy. The second important property refers to model completeness. It will be shown in theorem 5 that under some conditions, an NDAAO identified with a given paramter k simulates the complete original system language $L_{Orig}^{n \geq k}$: $L_{Ident}^{n \geq k} \supseteq L_{Orig}^{n \geq k}$.

For the following proofs it is advantageous to work with the identified NDAAO without merging of equivalent states. *Hence, it is assumed that the identified NDAAO was not given to the state merging algorithm.* Since in (Klein, 2005) it has been shown that merging of equivalent states does not affect the automaton language, the proofs for the automaton without merged states also hold for the automaton after algorithm 2. The following lemmas are necessary to proof theorem 4 and 5. As a first step we show that for each observed word of length k a state trajectory in the automaton exists which reproduces the observed word. The notation $x(i), \ldots, x(i + n - 1)$ denotes a sequence of states. If the output function λ is applied to the sequence, the result is the word of length n produced when the states of the sequence are passed: $\lambda(x(i), \ldots, x(i + n - 1)) = \lambda(x(i)), \ldots, \lambda(x(i + n - 1))$.

Lemma 1. In an NDAAO identified with parameter k, for each observed word of length $n \leq k$ a state trajectory exists which reproduces this word:

$$(\forall w^n \in L_{Obs}^k) \exists \lambda(x(i), \ldots, x(i + n - 1)) = w^n | (x(j + 1) \in f(x(j)) \forall i \leq j < i + n - 1)$$

In other words: The identified language simulates the observed language of length k: $L_{Ident}^k \supseteq L_{Obs}^k$.

For the proof the following definition is helpful:

Definition 22. The function $\widetilde{\lambda}(x)$ delivers the word of length k used in step 1 of algorithm 1 for the creation of the state. It represents the state output before it has been replaced by the last output symbol of the word in step 4. The function can only be applied to an NDAAO which has not been given to the state merging algorithm.

In the example of the former section, $\widetilde{\lambda}(x)$ delivers the state outputs depicted in figure 3.4 (e.g. AB for $\widetilde{\lambda}(x_1)$). The function $\widetilde{\lambda}(x)$ delivers a unique word for each state, since the automaton is considered without merging of equivalent states.

Proof of lemma 1. We proof that for each observed word of length k a state trajectory

$$(\forall w^k \in L_{Obs}^k) \exists \lambda(x(i), \ldots, x(i + k - 1)) = w^k | (x(j + 1) \in f(x(j)) \forall i \leq j < i + k - 1)$$

exists. The existence of shorter state trajectories $w^{n \leq k} \in L_{Ident}^k$ follows directly from the definition of the observed word set.

From equations 3.1 and 3.2 it follows that[6]

$$(\forall w^k \in L_{Obs}^k)(\exists v_1^k, \ldots, v_k^k \in W_{Obs, \Sigma^k}^k |$$
$$v_1^k \langle k \rangle = w^k \langle 1 \rangle, v_2^k \langle k - 1..k \rangle = w^k \langle 1..2 \rangle, \ldots, v_k^k \langle 1..k \rangle = w^k \langle 1..k \rangle$$

From steps 1 and 2 of algorithm 1 it follows that states representing $v_1^k, \ldots, v_k^k$ are connected such that

$$\widetilde{\lambda}(x(i)) = v_1^k,$$
$$\widetilde{\lambda}(x(i + 1)) = v_2^k \text{ and } x(i + 1) \in f(x(i)),$$
$$\ldots,$$
$$\widetilde{\lambda}(x(i + k - 1)) = v_k^k \text{ and } x(i + k - 1) \in f(x(i + k - 2))$$

---

[6]For the definition of the $\langle k \rangle$ operator (e.g. $v\langle k \rangle$) see equation 3.4 on page 44

since from equation 3.3 it follows that $(\forall v_i^k, v_{i+1}^k \in v_1^k, \ldots, v_k^k)(\exists w^{k+1} \in W_{Obs,\Sigma^k}^{k+1})|w^{k+1} = v_i^k\langle 1..k\rangle v_{i+1}^k\langle k\rangle$ (concatenation of the sequences $v_i^k\langle 1..k\rangle$ and $v_{i+1}^k\langle k\rangle$) and $(\forall v^k \in W_{Obs,\Sigma^k}^k)$ $(\exists!x)|\widetilde{\lambda}(x) = v^k$. Hence, there exists a trajectory which reproduces the observed word w^k when the state outputs are assigned to the last letter of the strings $v_1^k, \ldots, v_k^k$ (step 3 of the identification algorithm). $\qquad \square$

In the next lemma it is shown that each state x in the automaton is only reachable by a state trajectory of length $n \leq k$ if it produces a word $w^n = \widetilde{\lambda}(x)\langle k-n+1..k\rangle$ (the last n symbols of the word $w^k = \widetilde{\lambda}(x)$).

Lemma 2. We consider each state trajectory of length $n \leq k$ reaching state $x(i+n-1)$ in an NDAAO identified with parameter k:

$$(x(i), \ldots, x(i+n-1))|x(j+1) \in f(x(j))\forall i \leq j < i+n-1$$

The state $x(i+n-1)$ is only reachable by producing the word

$$\widetilde{\lambda}(x(i+n-1))\langle k-n+1..k\rangle = w^n \in L_{Ident}^k$$

Proof of lemma 2. From step 4 of the algorithm it follows that $w^n\langle n\rangle = \widetilde{\lambda}(x(i+n-1))\langle k\rangle$. From step 2 of the identification algorithm, it follows that $f(x(i+n-2))$ is only updated with $x(i+n-1)$ if $\widetilde{\lambda}(x(i+n-2))\langle 2..k\rangle = \widetilde{\lambda}(x(i+n-1))\langle 1..k-1\rangle$. This makes sure that $w^n\langle n-1\rangle = \widetilde{\lambda}(x(i+n-2))\langle k\rangle$. This principle can be continued: $f(x(i+n-3))$ is only updated with $x(i+n-2)$ if $\widetilde{\lambda}(x(i+n-3))\langle 2..k\rangle = \widetilde{\lambda}(x(i+n-2))\langle 1..k-1\rangle$. Hence, $w^n\langle n-3\rangle = \widetilde{\lambda}(x(i+n-3))\langle k\rangle$. This consideration can be repeated until the initial state $x(i)$ of the trajectory is reached with $w^n\langle k-n+1\rangle = \widetilde{\lambda}(x(i))\langle k\rangle$. This shows that state $x(i+n-1)$ is only reachable by state trajectories of length $n \leq k$ if they produce the word

$$\widetilde{\lambda}(x(i+n-1))\langle k-n+1..k\rangle = w^n \in L_{Ident}^k$$

$\qquad \square$

The next lemma says that an NDAAO state can only be reached by a trajectory producing a word of length $n \leq k$ that is part of the observed language. It is a continuation of lemma 2.

Lemma 3. We consider each state trajectory of length $n \leq k$ reaching state $x(i+n-1)$:

$$(x(i), \ldots, x(i+n-1))|x(j+1) \in f(x(j))\forall i \leq j < i+n-1$$

The state $x(i+n-1)$ is only reachable by producing the word

$$\widetilde{\lambda}(x(i+n-1))\langle k-n+1..k\rangle = w^n \in L_{Obs}^k$$

Proof of lemma 3. If a word $w^{n \leq k} \notin L_{Obs}^k$ with $\lambda(x(i), \ldots, x(i+n-1)) = w^{n \leq k} | x(j+1) \in f(x(j)) \forall i \leq j < i+n-1$ exists, it follows from lemma 2 that a state must exist with $\tilde{\lambda}(x(i+n-1))\langle k-n+1..k\rangle = w^{n \leq k} \notin L_{Obs}^k$. Step 1 of the identification algorithm makes sure that only states with $\tilde{\lambda}(x) = w^k \in W_{Obs, \Sigma^k}^k$ exist. Due to equation 3.2 W_{Obs, Σ^k}^k can be divided in two subsets: The first subset is W_{Obs}^k. From lemma 2 follows that states representing a word of this set can only be reached by producing a $w^{n \leq k} = \tilde{\lambda}(x(i+k-1))\langle k-n+1..k\rangle \in L_{Obs}^k$ because $(\forall w^k \in W_{Obs}^n)(\exists! x) | \tilde{\lambda}(x) = w^k$. The second subset of W_{Obs, Σ^k}^k consists of words resulting from duplicating the first symbol of the observed sequences up to $k-1$ times (equation 3.1): $W_{Dupl, \Sigma^k}^k = W_{Obs, \Sigma^k}^k \setminus W_{Obs}^k$. From assumption 1 and definition 20 it follows that $w^{k+1} \in W_{Obs, \Sigma^k}^{k+1}$ cannot end with $w^{k+1}\langle k-1\rangle = w^{k+1}\langle k\rangle$. Hence, there are no two successive states with $\lambda(x(i-1)) = \lambda(x(i))$. With lemma 2 it follows that a state with $\tilde{\lambda}(x) = w^k \in W_{Obs, \Sigma^k}^k$ can only be reached by a state trajectory of length $n \leq k$ producing the pairwise varying symbols $w^n = w^k\langle k-n+1\rangle, \ldots, w^k\langle k\rangle$ with $\tilde{\lambda}(x) = w^k$. It follows that the artificially introduced words in W_{Obs, Σ^k}^k resulting from duplicating the first observed output symbol up to $k-1$ times cannot be reproduced by the automaton. This leads to the conclusion that only words $w^{n \leq k} \in L_{Obs}^k$ can be produced. $\qquad \square$

Lemma 3 has important consequences for model accuracy and performance of the model as an observer in online diagnosis. As a first direct use, the lemma will be used to proof the next theorem:

Theorem 4 ($k+1$-completeness). *For a given value of the identification parameter k, the identified NDAAO is $(k+1)$-complete in the sense of (Moor et al., 1998), i.e. $\forall n \leq k+1$, $L_{Ident}^n = L_{Obs}^n$.*

Theorem 4 means that the NDAAO identified with a given value of the parameter k represents *exactly* the set of observed words of length lower of equal to $k+1$ and thus minimizes the exceeding language of length $k+1$. The following proof shows that this property holds for the automaton identified with algorithm 1.

Proof of theorem 4. For a given word $w^k \in L_{Obs}^k$, it follows from the proof of lemma 1 that a state trajectory exists with $\lambda(x(i), \ldots, x(i+k-1)) = w^k$. From lemma 3 it follows that the state $x(i+k-1)$ cannot be reached by any other word than the considered observed w^k. Due to step 2 of the identification algorithm, from state $x(i+k-1)$ it is only possible to reach a state $x(i+k)$ with $\tilde{\lambda}(x(i+k-1))\lambda(x(i+k)) = w^{k+1} \in L_{Obs}^{k+1}$ $(\tilde{\lambda}(x(i+k-1))\lambda(x(i+k))$ is the concatenation of the words $\tilde{\lambda}(x(i+k-1))$ and $\lambda(x(i+k)))$. Hence, the automaton can only produce words $w^{k+1} \in L_{Obs}^{k+1}$. $\qquad \square$

With the proof of lemma 4 follows that algorithm 1 delivers a model that is as accurate as the algorithm from (Klein, 2005) but is not restricted to system evolutions starting and ending with the same output symbol. This theorem shows that for a given k, the exceeding language L_{Exc}^k of the identified NDAAO can be eliminated. Like shown in section 2.4.1 this minimizes the number of non-detectable faults.

In the next theorem[7], it is stated that the identified language simulates the original system language of arbitrary length if the condition $L_{Orig}^{k+1} = L_{Obs}^{k+1}$ holds for a given value of the identification parameter k.

Theorem 5. If $L_{Orig}^{k+1} = L_{Obs}^{k+1}$, then $L_{Ident}^{k+n} \supseteq L_{Orig}^{k+n}$ with $n \geq 1$ for an NDAAO identified with parameter k.

Proof of theorem 5. $L_{Ident}^{k+1} \supseteq L_{Orig}^{k+1}$ since the identified NDAAO is $k + 1$ complete. For $k + 2$ it holds: $\forall w^{k+2} \in L_{Orig}^{k+2}$ a decomposition $a^k b^1 c^1 = d^1 e^1 f^k = s^1 u^k v^1 = w^{k+2} | a^k b^1, e^1 f^k, u^k v^1 \in L_{Orig}^{k+1} = L_{Obs}^{k+1}$ exists. Each state trajectory producing a^k ends in the same state $x(i)$ (lemma 3 and step 1 of algorithm 1: $(\forall a^k \in W_{Obs,\Sigma^k}^k)(\exists!x)|\widetilde{\lambda}(x) = a^k$). In step 2 of the algorithm, this state is connected with $x(i+1)|\widetilde{\lambda}(x(i+1)) = u^k$. Each state trajectory producing u^k ends in the same state $x(i+1)$ which gets connected to $x(i+2)|\widetilde{\lambda}(x(i+2)) = f^k$. Since there is a trajectory leading to state $x(i)$ and $x(i)$, $x(i+1)$ and $x(i+2)$ are in one trajectory, it follows that $\forall w^{k+2} \in L_{Orig}^{k+2}$ there exists a trajectory of states producing this word. It is obvious that for larger values than $k + 2$ $\forall w^{k+n} \in L_{Orig}^{k+n}$ there is always an appropriate decomposition into already observed substrings of $L_{Orig}^{k+1} = L_{Obs}^{k+1}$ to find a trajectory of connected states like presented above. Hence, it follows that $L_{Ident}^{k+n} \supseteq L_{Orig}^{k+n}$ if $L_{Orig}^{k+1} = L_{Obs}^{k+1}$ holds. $\square$

Based on the consideration from section 2.4.1, Theorem 5 shows that it is possible to eliminate the non-reproducible language $L_{NR}^{n \geq k+1}$ if it is possible to state $L_{Orig}^{k+1} = L_{Obs}^{k+1}$. This minimizes the number of false alerts during online fault diagnosis with the identified model. In the next section it is shown how it can be decided if $L_{Orig}^{k+1} = L_{Obs}^{k+1}$ holds for a given value of k.

3.4 Parameterization of the identification algorithm

3.4.1 Meaning of the identification tuning parameter

In this section new results concerning the role of the identification tuning parameter k are presented. It is shown how the identification parameter helps to distinguish non-equivalent states of a closed-loop DES leading to identical outputs.

In a closed-loop DES it is possible that different states lead to the same output symbol u^{DES} but have different fault-free following behaviors. As an example, figure 3.6 is considered. The figure shows a conveyor as typical part of a manufacturing system. The conveyor and its controller can be treated like a closed-loop DES. The closed-loop system of plant (conveyor) and controller exhibits the input/output (I/O) values of the controller as measurable system output u^{DES}. The current controller I/O vector consisting of the controller I/Os is thus considered as system output u^{DES} (a formal definition of the controller I/O vector as system output will be given in definition 24 in section 4.2.1). The conveyor has three position sensors P1, P2 and P3. The according controller-inputs are the first three elements of the I/O vector. If the position sensors detect a parcel, they return the value 1, else they return 0. The conveyor can be started

---

[7]This theorem was not shown for the algorithm of (Klein, 2005)

by setting an actuator to 1. Figure 3.6 shows the parcel in two different positions (position I and position II). It can be seen that the two positions lead to the same system output (C) but have a different fault-free following behavior: In position I, P2 is expected to change its value (leading to vector D) and in position II, P3 is expected to change its value (leading to vector E). The two underlying closed-loop DES states are thus not equivalent although they lead to the same output vector: According to (Cassandras and Lafortune, 2006), two states are only equivalent if they have the same output *and* if the language starting in them is equal.

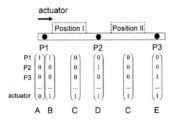

Figure 3.6: Conveyor example

An accurate model should be able to distinguish the two situations although they are represented by the same I/O vector. An automaton that can distinguish between the two non-equivalent system states is given in the right of figure 3.7. Position I corresponds to state x_2 and position II belongs to state x_4. If the two non-equivalent closed-loop DES states are *not* represented by two different states in an automaton like in the left part of figure 3.7, the following situation can arise: A fault lets the system perform the sequence BCE which is not part of the fault-free behavior. BCE is a fault symptom since the parcel must pass sensor P2 before it reaches P3 which is represented by I/O vector E. The automaton on the left has a state trajectory which leads to the generated word BCE (x_1, x_2 and x_4). The reason is that x_2 can produce the following behavior of both underlying non-equivalent closed-loop DES states. BCE is part of the exceeding behavior which makes the detection of the considered fault impossible with the automaton in the left of figure 3.7. If a model in form of an automaton is built manually, an intuitive way to achieve high accuracy is to represent only *equivalent* DES states with one automaton state. Two automaton states are intuitively connected if the represented DES states can occur successively.

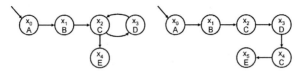

Figure 3.7: Automata for the conveyor example

If a model is built by identification, the decision if two identical output symbols u^{DES} belong to two different closed-loop DES states must be made based on an analysis of the observed system data. In the example in figure 3.6 it can be seen that using the preceding I/O vector of position I and II allows distinguishing the two underlying closed-loop DES states: If vector C occurs due to position I, it must be the successor of vector B. If C is caused by position II, it must follow vector D.

In the identification algorithm from section 3.5.3, the parameter k allows integrating knowledge concerning the preceding system outputs in an NDAAO state. In lemma 3 it is shown that an NDAAO state is only reachable by a state trajectory producing an output sequence that corresponds to the word of length $n \leq k$ given by the concerning state output $\widetilde{\lambda}(x)$. In the identification algorithm, two NDAAO states are only connected if they share the same memory of length k. If k is chosen such that it is possible to distinguish two non-equivalent closed-loop DES states with the same output by their preceding sequences of length k, the identification algorithm does not represent them with the same automaton state. In this case it is made sure that each automaton state can only produce the fault-free following behavior of *one* closed-loop DES state. Any other possibly faulty following behavior cannot be produced. This minimizes the number of non-detectable faults during online fault diagnosis. In the example of figure 3.6 it is sufficient to chose $k = 2$ since the non-equivalent closed-loop DES states with the same output are distinguishable by considering sequences of length two.

3.4.2 Discussion on an upper bound for the identification parameter

In section 2.4.1 it has been explained that the performance of a model in online diagnosis is determined by the minimization of the exceeding and the non-reproducible behavior. It has been explained that the exceeding behavior refers to non-detectable faults and the non-reproducible behavior corresponds to false alerts. Choosing a large value for k improves the distinction of non-equivalent closed-loop DES states by a model state which consequently leads to a reduced exceeding behavior. In this section it is shown that in most practical cases this reduction comes at cost of an increased non-reproducible behavior which is closely related to the number of false alerts: a word of the non-reproducible language is part of the normal system behavior but it cannot be reproduced by the (fault-free) model during online diagnosis (see section 2.4.1). Hence, a false alert is risen. It will be concluded that in practical applications, an upper bound for the identification parameter k exists.

In (Blanke et al., 2006) it is shown that for fault diagnosis it is crucial to have a *complete* model of the considered system. In terms of language theory, a model is complete if its language simulates the fault-free language of the considered system. In theorem 5 it has been shown that the identified language L_{Ident}^{k+n} simulates the original system language L_{Orig}^{k+n} for $n \geq k$ if $L_{Obs}^{k+1} = L_{Orig}^{k+1}$ holds. The following considerations will show that for large values of k it is often not possible to state $L_{Obs}^{k+1} = L_{Orig}^{k+1}$.

Figure 3.8 shows a typical evolution of the observed language of a closed-loop system. The data is taken from a case study which is treated in section 3.5. It is shown how the set cardinality of the observed language L_{Obs}^n evolves after several observed system evolutions. It can be seen that the cardinality of the language of length $n = 1$ converges to a stable level after 10 observed system evolutions. From the graph it can also be seen that L^2_{Obs} takes longer than L^1_{Obs} to converge to a stable level. Until the 15th system evolution, the curve representing $|L^2_{Obs}|$ has a positive gradient. Generally, it can be observed that the set cardinality of L^n_{Obs} for larger values of n takes longer to converge to a stable level than in the case of smaller n. The evolution of the observed system language is very similar to the evolution sketched in figure 3.3 on page 43 which was shown to illustrate assumption 3 stating that the observed system language converges to the original language with growing observation time.

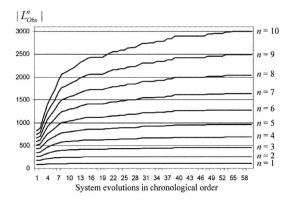

Figure 3.8: Typical evolution of the observed language

If assumption 3 holds, the following consideration can be made: If after a phase of growth, $|L^n_{Obs}|$ is not augmented for a sufficiently long time, it is a reasonable heuristic to assume that $|L^n_{Obs}|$ will not grow in the future. If L^n_{Obs} does no longer grow, it has converged to L^n_{Orig}: The system has exhibited each $w^n \in L^n_{Orig}$ which is possible during normal operation.

Assumption 5 (Complete observation). If the set cardinality of L^n_{Obs} strongly converges to a stable level for a *significant number* of system evolutions, it is assumed that $L^n_{Obs} \approx L^n_{Orig}$.

The *significant number* of system evolutions mentioned in the assumption is a parameter of the representativeness of the observed data sample. It must be decided by the user if the significant number is reached.

If an identified NDAAO is to be used for online fault diagnosis, the parameter k must be chosen such that $L^{k+1}_{Obs} \approx L^{k+1}_{Orig}$ to get a complete model simulating the original system language of length $n \geq k + 1$ (theorem 5). If the NDAAO is identified on an incomplete data basis $L^{k+1}_{Obs} \subseteq L^{k+1}_{Orig}$, it is likely that the system exhibits new words $w^{k+1} \in L^{k+1}_{Orig}$ that have not yet been observed and are thus not part of the identified language since from theorem 4 it follows that $L^{k+1}_{Ident} = L^{k+1}_{Obs}$. These new fault-free words are not reproducible by the identified model and thus lead to erroneous fault detection. Each not yet observed fault-free word $w^{k+1} \in L^{k+1}_{Orig}$ leads to fault detection since it is not part of the model language. It follows that it is not possible to chose a too large value for the identification parameter k since this leads to a high number of false alerts. k can only be chosen to a value such that assumption 5 holds for $L_{Obs}^{n=k+1}$. In figure 3.8 the convergence criterion allows for example stating $L_{Obs}^{n=3} \approx L_{Orig}^{n=3}$. In this case, k can be chosen to $k = n - 1 = 2$.

3.4.3 Discussion on a lower bound of the identification parameter

In the former section it has been shown that the upper bound for the identification parameter k is determined by the convergence of the observed language. Generally, it can be said that a small value for k delivers a model leading to a small number of false alerts whereas a large value for k increases the sensitiveness to faults (reduction of the exceeding language) but also leads to a higher tendency for false alerts. In figure 3.8 it can be seen that for larger values of n, it may also be possible to state that L_{Obs}^{n} is rather completely observed although the convergence for lower values is clearer. Hence, in some cases it is possible to choose a relatively large value for k which still leads to a rather complete data base for identification. To judge if the higher tendency for false alerts when increasing k is acceptable since the model accuracy is significantly increased, a measure for the positive effects of increasing k is introduced.

In section 3.4.1 it is explained that an NDAAO state x can represent several closed-loop DES states $\{x_1^{DES}, x_2^{DES}, \ldots, x_i^{DES}\}$ with different valid following states. If k is chosen such that the model is complete (it can reproduce the complete fault-free original system language), the NDAAO state leads to states representing the successors of $\{x_1^{DES}, x_2^{DES}, \ldots, x_i^{DES}\}$. Hence, the transition function of the NDAAO state x yields more states than each single transition function of one of the represented DES states. If the considered system is in state x_1^{DES} and a fault occurs, it is possible that this fault leads to a following behavior which is valid in another system state x_2^{DES} but not in x_1^{DES}. Since the NDAAO represents the system being in x_1^{DES} with the same state that represents x_2^{DES}, the faulty behavior can be reproduced. The NDAAO state cannot distinguish between the system states x_1^{DES} and x_2^{DES}. Hence, if a following behavior is valid in x_2^{DES}, it is also supposed to be valid in x_1^{DES}.

This consideration shows that if by increasing k it is possible to make the approximation of DES states by NDAAO states less ambiguous, the fault detection capability of the identified model can be improved. As an example of this effect, the identified automata in figure 3.9 are considered. The data basis for the identification is given by the following two observed sequences. They are supposed to be the only possible fault-free sequences of the closed-loop DES:

$$\sigma_1 = (A, B, C, D, B, C, D, E, F, G, H)$$
$$\sigma_2 = (A, B, E, F, G, H)$$

with

$$\sigma_1 = \Lambda((x_0^{DES}{}^A, x_1^{DES}{}^B, x_2^{DES}{}^C, x_3^{DES}{}^D, x_4^{DES}{}^B, x_5^{DES}{}^C, x_6^{DES}{}^D, x_7^{DES}{}^E, x_8^{DES}{}^F, x_9^{DES}{}^G, x_{10}^{DES}{}^H))$$

$$\sigma_2 = \Lambda((x_0^{DES}{}^A, x_1^{DES}{}^B, x_{11}^{DES}{}^E, x_{12}^{DES}{}^F, x_{13}^{DES}{}^G, x_{14}^{DES}{}^H))$$

denoting the underlying closed-loop DES states of the observed sequences. On top of each state its output $\Lambda(x^{DES})$ is written. It can be seen that $\Lambda(x_1^{DES}) = \Lambda(x_4^{DES}) = B$: x_1^{DES} and x_4^{DES} have the same output but are not equivalent since their following behavior differs: In contrast to x_1^{DES}, from x_4^{DES} it is not possible to produce $BCDB$.

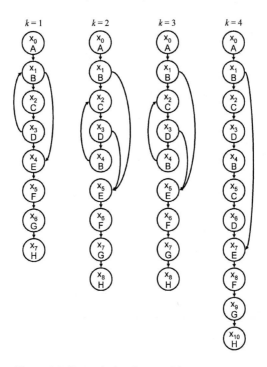

Figure 3.9: Example for the transition gap measure

In figure 3.9, the identified automata for different values of k are given. The automaton structure varies when increasing k. If the observed system is in state x_3^{DES} with $\Lambda(x_3^{DES}) = D$, the automaton at $k = 1$ represents this situation with x_3. From state x_3^{DES}, the only valid following behavior leads to the observation of B (see σ_1). If a fault occurs leading to a non-valid following behavior E, this can be reproduced by the automaton with the actual state x_3. It cannot distinguish between the DES states x_3^{DES} and x_6^{DES} (with $\Lambda(x_3^{DES}) = \Lambda(x_6^{DES}) = D$). Since from x_6^{DES}, E is a valid following behavior, the fault cannot be detected due to its behavior being part of the model.

Figure 3.9 shows that increasing k to $k = 2$ changes the automaton structure. It can be seen that the automaton identified with $k = 1$ contains an exceeding word $w^3 = DBE$ which by theorem 4 is not part of the automaton identified with $k = 2$ (since $L_{Ident}^{k+1} = L_{Obs}^{k+1}$). Although increasing k to $k = 2$ thus improves the overall accuracy, the automaton identified with $k = 2$ still contains a state representing two different non-equivalent system states: x_3 still models x_3^{DES} and x_6^{DES}. Hence, a fault leading to the output E from x_6^{DES} can still not be detected. Increasing k to $k = 3$ does not change this situation since the algorithm yields the same automaton structure. Only if k is chosen to $k = 4$, the automaton state x_3 models only x_3^{DES} since it is possible to distinguish between x_3^{DES} and x_6^{DES} by considering their three predecessors. With $k = 4$, x_3 has only one possible following state. Hence, the fault leading to E from x_3^{DES} can be detected, because E is no longer an accepted following behavior of x_3[8]. A closer look at the example shows that not only x_3 representing x_3^{DES} but the whole trajectory x_2, x_3 representing x_2^{DES}, x_3^{DES} and x_5^{DES}, x_6^{DES} was duplicated such that the new trajectories at $k = 4$ represent only one of the underlying DES-trajectories.

If NDAAO trajectories represent less DES trajectories, this leads generally to a lower number of states with multiple leaving transitions. Since the duplicated NDAAO trajectory does no longer have to represent the different following behaviors of two DES trajectories (to assure the completeness of the model), the number of leaving transitions of the last state in the two new NDAAO trajectories is reduced. If this effect takes place, the number of transitions is thus not as quickly augmented as the number of states when increasing k. In the following definition, the difference of states and transitions of an NDAAO is introduced.

Definition 23 (Transition-state gap). The transition-state gap of an NDAAO identified with parameter k is defined as

$$\text{Gap}(NDAAO, k) = |f(X)| - |X|$$

It is the difference of the number of transitions (denoted as $|f(X)|$) and the number of states ($|X|$) of an NDAAO identified with parameter k.

In table 3.1 the evolutions of the number of states and transitions and of the Gap-function is depicted for the example. Increasing k from $k = 1$ to $k = 3$ adds the same number of states and transitions. If k is chosen to $k = 4$, more states than transitions are added which is a strong indicator that an NDAAO trajectory representing two DES trajectories at $k = 3$ has been split to two NDAAO trajectories. Hence, the Gap-function is decreased at $k = 4$.

The preceding considerations show that a decreased Gap-function is a strong indicator that the representation of DES trajectories has become less ambiguous by increasing k. It is a possibility to get information about the representation of non-equivalent DES states *although the original DES states are not known in the case of black-box identification*. Hence, the following heuristic set of rules to determine an appropriate

---

[8]This holds if an appropriate state estimation algorithm is used, delivering x_3 as unique estimate. Such an algorithm is introduced in chapter 6.1

k	$|X|$	$|f(x)|$	Gap($NDAAO, k$)
1	8	9	1
2	9	10	1
3	9	10	1
4	11	11	0

Table 3.1: Number of states and transitions

value for k is proposed if k cannot be chosen upon a priori knowledge. Since the rules make use of the transition-state gap, it is necessary to identify models with different values for k before they can be applied. The rules have thus to be understood as an a posteriori heuristic.

Rule 1: If it is not possible to determine a value $k \geq 1$ such that $L_{Obs}^{k+1} \approx L_{Orig}^{k+1}$ can be stated due to the convergence of L_{Obs}^{k+1} to a stable level, it is necessary to increase the observation horizon and to collect more data. Each value $k \geq 1$ for which $L_{Obs}^{k+1} \approx L_{Orig}^{k+1}$ holds is an appropriate value to identify a model eliminating the non-reproducible system language. This leads to a minimum number of false alerts if the model is used for online diagnosis purposes.

Rule 2: If $L_{Obs}^{k+1} \approx L_{Orig}^{k+1}$ holds for several k but with lower confidence if k is increased, k should only be augmented if the transition-state gap decreases significantly.

Choosing k with these rules is a compromise between a low number of false alerts and high model accuracy which leads to an increased fault detection capability.

3.5 Case study: Fischertechnik laboratory facility

3.5.1 System description

To demonstrate the impact of the proposed methods for existing closed-loop DES, a case study has been treated. It will be used as a running example in the following chapters. The considered system is a laboratory facility at the institute of automatic control, University of Kaiserslautern. It has typical characteristics of an industrial manufacturing process. The purpose of the system depicted in figure 3.10 is to treat work pieces that are stored in the feeder (left most station). The system is controlled using a Siemens S300 PLC (Programmable Logic Controller) with 15 inputs and 15 outputs. The controller inputs and the corresponding sensors can be seen in figure 3.10. Inputs written in *italic* are connected to sensors with a specific technology that delivers a logical 0 if they detect something and a logical 1 if they do not detect anything. The outputs are given in table 3.2. If they are set to 1, the according actuator gets activated. The I/O names are labeled with I1.2 etc. to indicate the second *i*nput that belongs to the first machine and O2.4 for the fourth *o*utput of the second station.

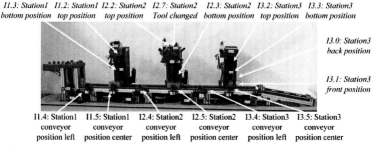

Figure 3.10: Laboratory system

A system evolution consists of the treatment of a given number of work pieces that are stored in the feeder. As a first scenario, it is supposed that two work pieces are to be treated. At the beginning of a system evolution the feeder pushes the first work piece to the conveyor. Then it is transported to the drilling machine (station 1). In the drilling station the work piece is stopped and two holes are drilled. As soon as the first work piece has been treated in the first station it is transported to the vertical milling machine (station 2) and the next piece is taken from the feeder. The first work piece gets treated by the vertical milling machine (each of the three milling tools is applied to the piece successively) and the second one by the drilling station. After the treatment at the vertical milling machine has been finished, the work piece gets transported to the horizontal milling station (station 3). The second work piece moves from station 1 to station 2. When the first work piece has been treated in station 3, it is stored in the last station at the right side in figure 3.10. This process continues until both work pieces have been treated by each of the three stations.

3.5.2 Data collection for identification and online diagnosis

As explained in section 3.1, the proposed identification method works on the basis of signals exchanged between controller and plant. The data collection method chosen for the case study works with a standard communication processor that is part of the PLC. When controlling the system evolution, the PLC performs the following three steps cyclically: reading the inputs, executing the program in order to determine the new output setting and finally writing the outputs. As depicted in figure 3.11, the transfer of the newly determined I/O values in form of an I/O vector takes place at the end of the second step (program execution). At the same time when the newly determined outputs get transfered to the output card of the PLC, the new I/O vector is sent to a standard PC using the communication processor via a UDP-connection. The languages defined in section 3.1 are thus based on an alphabet which consists of the set of I/O vectors exhibited by the system. Since the I/O vectors are captured at the end of the PLC cycle and are thus sampled, it is possible to have multiple I/Os changing their value between two I/O vectors. This method can be applied to each

I/O	Description	I/O	Description
O1.2	drilling machine (station 1) motor up	O2.2	vertical milling machine (station 2) motor up
O1.3	drilling machine motor down	O2.3	vertical milling machine motor down
O1.4	drilling machine drilling motor	O2.4	vertical milling machine milling motor
O1.5	drilling machine conveyor on	O2.5	vertical milling machine conveyor on
		O2.7	change milling head
O3.0	horizontal milling machine (station 3) motor back	O3.1	horizontal milling machine motor front
O3.2	horizontal milling machine motor up	O3.3	horizontal milling machine motor down
O3.4	horizontal milling machine milling motor	O3.5	horizontal milling machine conveyor on

Table 3.2: Controller outputs of the case study

system with a controller that is able to send data via an Ethernet data link which is a common feature of almost all industrial controllers today. A detailed description of the data collection procedure and its consequences on the collected data is given in (Roth et al., 2010).

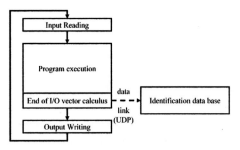

Figure 3.11: PLC cycle with data collection

3.5.3 Identification of monolithic models

In figure 3.8 on page 54, the evolution of the observed language L_{Obs}^n for the case study for n up to $n = 10$ has already been shown. In the former sections it has been explained that for identification of an NDAAO, k should be chosen such that $L_{Obs}^n \approx L_{Orig}^n$ with $n = k + 1$ which by assumption 5 is expressed by the convergence of $|L_{Obs}^n|$ to a stable level. It can be seen that for lower values of n, the language converges very quickly and can thus reliably be considered as completely observed. For larger values of n, the convergence can also be observed, but cannot as reliably be stated as in the case of a lower n. In order to decide if the higher tendency for false alerts induced by increasing k is acceptable due to an increased fault detection capability of the model, in figure 3.12, the Gap-function of definition 23 is depicted for several values of k. It can be seen that the function decreases significantly when k is increased from $k = 1$ to $k = 2$. Hence, this is a strong indicator that the model accuracy is improved since the number of ambiguous DES state representations is decreased. It can be seen that at $k = 4$, the Gap-function decreases again. Since this decrease is not very important, the quality of the model is not significantly improved by increasing k to $k = 4$. Increasing k beyond $k = 4$ does not lead to a falling gap-function. Based on the available data and the evolution of the Gap-function, it is thus a reasonable decision to chose $k = 2$ for the identification algorithm.

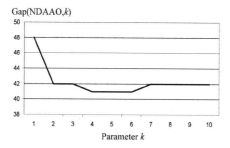

Figure 3.12: Evolution of the Gap-function

The automaton identified with $k = 2$ has 121 states and 163 transitions. Parts of automata identified with $k = 1$ and $k = 2$ can be seen in figure 3.13. Since the whole I/O vector is too large to be depicted in each state, only the I/Os changing their value when taking a given transition are shown. The complete state output can be reconstructed from the knowledge of the output of x_0 in each automaton: In x_0, all I/Os have the value 0 except of the following ones: I1.3, I2.3, I3.1 and I3.3 are 1. The notation of changing I/O values in figure 3.13 is formalized in chapter 6.1. Basically, I1.4_1 O1.5_1 between x_0 and x_1 means that I1.4 and I1.5 change from 0 to 1. Due to the data collection procedure explained in section 3.5.2, several I/Os can change their value from one vector to a following vector. I1.4_0 means that the corresponding input changes its value from 1 to 0.

Both automata in figure 3.13 represent the start of the system evolution. First, the automaton identified with $k = 1$ is considered. From x_0 to x_1 the first work piece is pushed on the conveyor and gets transported to station 1 (x_2). Then a first hole is drilled into the work piece. After this, the work piece is moved on until the position sensor I1.5 changes its value (x_8). The conveyor stops again and a second hole is drilled (x_8 to x_{11}). When the tool is back at its top position, the work piece is moved on to the next station, which is represented by x_2. Here it can be seen that the identified automaton cannot distinguish between the work piece being in the middle of I1.4, I1.5

and between the work piece being in the middle of I1.5 and I2.4. This represents a model inaccuracy. An exceeding behavior of length three can be produced when the automaton models the work piece moving from I1.4 to I1.5 but arriving at I2.4 instead of I1.5 (state trajectory x_1, x_2, x_{12} instead of trajectory x_1, x_2, x_3). If a fault occurs leading to the I/O vector of x_{12}, when the work piece is between I1.4 and I1.5, it cannot be detected.

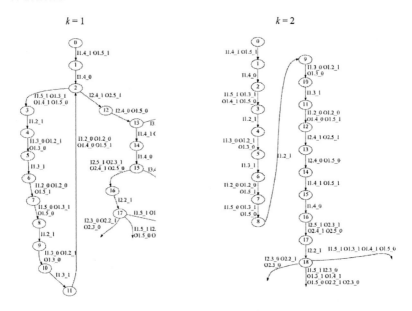

Figure 3.13: Parts of identified automata for the case study

This inaccuracy can be omitted by increasing k to $k = 2$. Where the automaton identified with $k = 1$ loops back to x_2, the more accurate NDAAO at $k = 2$ represents the work piece being between I1.5 and I2.4 with an own state (x_{12}). At $k = 2$ the non-equivalent system states describing the work piece between I1.4 and I1.5 on the one hand and I1.5 and I2.4 on the other hand are distinguishable by analyzing their predecessors. Since several other states are also split from $k = 1$ to $k = 2$, this explains the falling Gap-function in figure 3.12. This shows that choosing $k = 2$ leads to model which is significantly more accurate than at $k = 1$ with a data base that can reasonably be considered as completely observed.

4 Distributed Identification of Closed-Loop Discrete Event Systems

4.1 Limits of monolithic identification

In the former chapter it has been shown that an important precondition for the identification of a closed-loop DES is to state that the original system language can be considered as completely observed. In assumption 5 it has been explained that for this purpose, the convergence of the observed language to a stable level is a suitable indicator. In figure 3.8 the evolution of the observed system language for the case study from section 3.5 has been shown. The observed system evolutions consist of the treatment of *two* work pieces which led to the convergence of the observed language L_{Obs}^n even for relatively large values of n. If the number of work pieces to be treated is increased to *three*, the observed system language does not converge as quickly as with *two* work pieces.

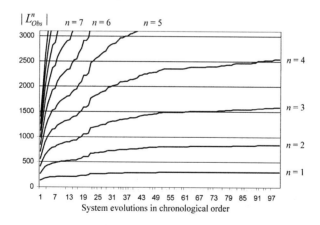

Figure 4.1: Observed language of the case study with three work pieces

Figure 4.1 shows the evolution of the set cardinality of L_{Obs}^n for this case. Although more system evolutions than in the case of treating two work pieces have been observed (100 versus 60 evolutions), only $|L_{Obs}^1|$ converges to a stable level. This convergence takes 50 evolutions whereas the convergence of $|L_{Obs}^1|$ with two work pieces only took 10 evolutions. In the former chapter it has been explained that the convergence of the observed system language to a stable level is crucial to get a complete model. Without a complete model simulating the whole fault-free system language the number of false alerts during online fault diagnosis can become unacceptably high. The reason for the longer growing phase of $|L^1_{Obs}|$ is the higher degree of concurrency induced by increasing the number of work pieces to be treated.

If three work pieces are to be treated in the case study system, the three stations depicted in figure 3.10 work simultaneously when the third work piece has been pushed on the first conveyor and the first two work pieces are treated by the second and the third station. Like in the case of Petri nets as shown in figure 4.2, a part of the behavior is not synchronized and can be executed in different orders. First, the leftmost Petri net is considered. It consists of two parallel branches. After the occurrence of event a, the following two sequences are possible:

$$s_1 = abcde \qquad\qquad s_2 = acbde$$

In this case, the events b and c can be generated concurrently. Event d can only occur when b and c both have occurred. If concurrency is increased like in the rightmost Petri net in figure 4.2, the number of possible generated event sequences is increased. The Petri net with three parallel branches can generate the following six evolutions:

$$s_1 = abcde \qquad s_2 = abdce \qquad s_3 = acbde$$
$$s_4 = acdbe \qquad s_5 = adbce \qquad s_6 = adcbe$$

In this case the events b, c and d can be generated concurrently after the occurrence of a. This example shows that increasing concurrency in the system increases the original system language.

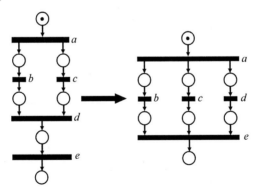

Figure 4.2: Example for concurrent behavior

In the case study, the simultaneous treatment of three work pieces in three stations also increases concurrency: After a work piece has arrived at a station, the station behavior is independent from the behavior of the other stations. Like in the Petri net example, the combined behavior of such concurrent subsystems (represented by individual branches in the Petri net) leads to a large number of possible global system behaviors. If the global behavior of such concurrent systems is considered, it is a direct consequence that it takes the system longer to exhibit a significant part of its original system language since the size of the original language increases with augmenting concurrency. Hence, the observed language takes longer to converge to a stable level as depicted in figure 4.1.

A possible solution for the problem of a non-converging observed language is to consider *subsystems* with a lower degree of internal concurrency. In the rightmost Petri net in figure 4.2 such a subsystem could consist of two of the three parallel branches. For the case study system, a subsystem could consist of the two left most stations. Figure 4.3 shows the evolution of the observed language if only the first two stations are considered. Considering only two stations, the maximum number of work pieces treated in parallel is two. It can be seen that the concerning observed system language converges significantly faster than in figure 4.1 where the complete system is considered. Since the model for the first two stations can be identified on the basis of a converged observed language $L_{Obs}^{k+1} \approx L_{Orig}^{k+1}$ (e.g. with $k = 2$), the precondition of theorem 5 is fulfilled which leads to a minimized number of false alerts during online fault diagnosis.

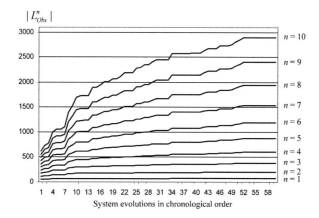

Figure 4.3: Observed language of a subsystem consisting of the first two stations treating three work pieces

In many practical applications, the same phenomenon as in the case study system treating three work pieces can be observed: The observed system language does not converge to a stable level, which does not allow identifying a suitable model for online fault diagnosis. To overcome this problem, it is possible to divide the system in subsystems with converging observed languages. The question of how to choose appropriate subsystems is treated in chapter 5. In the remaining part of this chapter an approach is proposed to identify subsystem models of a closed-loop DES. It will be shown that fulfilling the precondition $L_{Obs}^{k+1} \approx L_{Orig}^{k+1}$ for each single subsystem comes at cost of an increased global exceeding language which is a disadvantage for online diagnosis. To limit the global exceeding system behavior, an approach to systematically restrict the combined behavior of an identified automata network is presented.

4.2 Identification of partial closed-loop DES

4.2.1 Definition of partial closed-loop DES

As explained in section 3.5, the data base for identification is determined by collecting the signals exchanged between controller and plant. These signals can be grouped in an I/O vector:

Definition 24 (Controller I/O vector). Given r different controller inputs I_1, ..., I_r and s different controller outputs O_1, ..., O_s, the controller I/O vector $u = (IO_1, ..., IO_m)$ with $m = r + s$ is given by $IO_i = I_i \ \forall \ i = 1, .., r$ and $IO_{r+i} = O_i \ \forall \ i = 1, .., s$. $m = |u|$ denotes the length of the vector (number of controller I/Os). The controller I/Os can have the values 1 or 0.

In the following, the controller I/O vector u is considered as the closed-loop DES output u^{DES} introduced in section 3.1. The controller I/O vector is the output of the complete system. If the system is divided into subsystems, only parts of the complete I/O vector are considered in each subsystem. To determine the I/Os which are to be considered in a subsystem, a mapping function is introduced:

Definition 25 (Subsystems and I/O-mapping function). The closed-loop DES consists of several subsystems sys_t. The function $y(sys_t)$ assigns a set of controller I/Os to each subsystem. The number of subsystems is defined as N_{sys}.

The output of the t-th subsystem sys_t is determined by the partial I/O vector which contains the I/Os from the set $y(sys_t)$.

Definition 26 (Partial controller I/O vector). Given the controller I/O vector u according to definition 24: The *partial* controller I/O vector is defined as $u_{sys_t} = (IO_1, ..., IO_m)$ with $IO_i = -$ (don't care symbol) if $IO_i \notin y(sys_t)$ and IO_i taken from u if $IO_i \in y(sys_t)$.

Figure 4.4 shows the principle of the previous definitions. The global I/O vector of the complete system consists of five controller I/Os. The function $y(sys_1)$ assigns IO_2, IO_3 and IO_5 to the first subsystem whereas IO_1, IO_3 and IO_4 are assigned to the second subsystem. It can be seen that an I/O can be part of several subsystems. I/Os which are not part of the considered subsystem are replaced by the don't care symbol '-'.

Like the definition of the global observed system language L^n_{Obs} on the basis of the observed system output u, it is possible to define the *partial* observed system language based on the partial system output u_{sys_t}.

Definition 27 (Observed partial language). The observed partial system language of the t-th subsystem L^n_{Obs,sys_t} is defined according to definition 20 considering the partial I/O vector u_{sys_t} as observed system output.

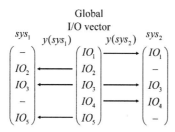

Figure 4.4: Division of the global I/O vector into two subsytems

Remark 2. Corresponding to assumption 1, it is assumed that the j-th partial I/O vector $u_{sys_t}(j)$ is created by the j-th event in the underlying subsystem. Hence, $u_{sys_t}(j) \neq u_{sys_t}(j+1)$ also holds for the partial system language.

On the basis of the observed partial system language L_{Obs,sys_t}^{k+1} a partial NDAAO$_t$ for the t-th subsystem is identified with the approach of chapter 3. It is obvious that the NDAAO$_t$ is $k+1$-complete with respect to L_{Obs,sys_t}^{k+1}. If the subsystems are chosen such that L_{Obs,sys_t}^{k+1} converges to a stable level, the identified partial automaton is an appropriate fault detection model to minimize the number of false alerts when observing the considered subsystem. In the next section the combined behavior of several NDAAO$_t$ representing several subsystems sys_t will be derived. This enables an evaluation of the fault detection capability of a set of partial automata representing the subsystems of a given global system.

4.2.2 Composition of partial closed-loop DES

After the identification of a partial automaton for each subsystem, the combined automaton behavior represents the identified language L_{Ident}^n. In figure 4.5 two identified NDAAO are shown (the function J is introduced in definition 28). They build an automata *network* representing the considered closed-loop DES. Each automaton has a partial I/O vector as state output and is able to reproduce an identified partial language L_{Ident,sys_t}^n. To determine the language of the automata network, the partial automata must synchronously perform state trajectories. The resulting outputs consisting of partial I/O vectors can then be combined to the global model output. To implement the synchronization of several NDAAO$_t$, the NDAAO cross product is introduced in this section. It is an adaptation of the parallel composition from (Cassandras and Lafortune, 2006) to the automaton of type NDAAO given in definition 16.

A first step for the construction of the NDAAO cross product is the following definition. It defines the result of combining two partial I/O vectors from different subsystems.

Definition 28 (Join function for two partial I/O vectors). Given two partial I/O vectors u_{sys_1} and u_{sys_2} with the same number of I/Os ($|u_{sys_1}| = |u_{sys_2}|$). The join function

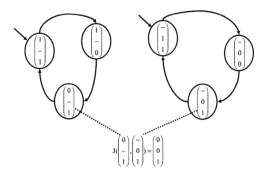

Figure 4.5: Two partial NDAAOs with a combined output

$J\left(u_{sys_1}, u_{sys_2}\right)$ delivers a vector with $|u_{sys_1}|$ elements. The i-th vector element is addressed with $J\left(u_{sys_1}, u_{sys_2}\right)[i]$:

$$J\left(u_{sys_1}, u_{sys_2}\right)[i] = \begin{cases} u_{sys_1}[i] & \text{if } u_{sys_1}[i] = u_{sys_2}[i] \\ u_{sys_2}[i] & \text{if } u_{sys_1}[i] = -\wedge u_{sys_2}[i] \neq - \\ u_{sys_1}[i] & \text{if } u_{sys_1}[i] \neq -\wedge u_{sys_2}[i] = - \\ c & \text{if } - \neq u_{sys_1}[i] \neq u_{sys_2}[i] \neq - \end{cases}$$

$\forall i = 1, \ldots, |u_{sys_1}|$, with $u_{sys_1}[i]$ denoting the i-th IO of the partial vector u_{sys_1}.

The I/Os at each position of the two partial I/O vectors are compared. If one of them is overwritten by the don't care symbol '-' since it does not belong the the according subsystem, the value of the other vector is taken. If one of the compared I/Os is 1 and the other one is 0, c for contradiction is written at the position of the according I/O. For the join function the associative law holds:

$$J(\lambda(x_{sys_1}), \lambda(x_{sys_2}), \lambda(x_{sys_3})) = J(\lambda(x_{sys_1}), J(\lambda(x_{sys_2}), \lambda(x_{sys_3}))) \tag{4.1}$$

In figure 4.5, the join function of two state outputs is given as an example. From this definition it follows directly that a valid state combination $\{x_{sys_1}, \ldots, x_{sys_n}\}$ of n partial automata modeling a system state must fulfill the following condition with m denoting the dimension of the I/O vector:

$$J(\lambda(x_{sys_1}), \ldots, \lambda(x_{sys_n}))[i] \neq c \quad \forall 1 \leq i \leq m \tag{4.2}$$

The reason is that the output of the considered closed-loop DES cannot contain the contradiction symbol c.

The language of an automata network is determined by the combined evolution of the underlying automata. In (Cassandras and Lafortune, 2006), the parallel composition is introduced as an appropriate operation to determine the combined behavior of interacting subsystems with *private* and *common* events. Private events are only related to one subsystem whereas common events are shared by several subsystems. The parallel composition of automata synchronizes common events. A common event can only be executed if the underlying automata execute it synchronously. Private events can be executed whenever possible. Subsystems with partial I/O vectors as system output are a typical example for systems with private and common events. I/Os considered in two subsystems are related to common events. If such an I/O changes its value in one subsystem, this must also happen in the second subsystem such that a valid system output according to equation 4.2 is possible. I/Os considered in one subsystem only, refer to private events. These I/Os can change their value independently since they are replaced by the don't care symbol '-' in the other subsystems. In definition 9 section 2.3.2, the parallel composition has been introduced for the 'standard'-automaton used in the diagnoser approach. In the following, an adaptation of this composition procedure to automata of the NDAAO-type is given. The operation will be denoted as 'automata cross product' or 'parallel composition' in the rest of the work.

The composition of two partial $NDAAO_1$ and $NDAAO_2$ is given in the following definition:

Definition 29 (Cross product or parallel composition of two partial NDAAO). The cross product

$$NDAAO_{||} = NDAAO_1 || NDAAO_2 = (X_{||}, J(\Omega_1, \Omega_2), f_{||}, \lambda_{||}, x_{||_0})$$

of two partial $NDAAO_1$ and $NDAAO_2$ with output alphabets Ω_1 and Ω_2 consisting of partial I/O vectors of the same length is an NDAAO according to definition 16. The single elements are given by the following equations:

State space:

$$X_{||} := \{(x_1, x_2) \in X_1 \times X_2 | c \notin J(\lambda(x_1), \lambda(x_2))\} \tag{4.3}$$

State output:

$$\forall (x_1, x_2) \in X_{||} : \lambda_{||}((x_1, x_2)) := J(\lambda(x_1), \lambda(x_2)) \tag{4.4}$$

Transition function:

$$\forall (x_1, x_2) \in X_{||} : f_{||}((x_1, x_2)) := \{(x_1', x_2') \in X_{||} | (x_1', x_2') \in \{x_1 \cup f_1(x_1)\} \times$$
$$\{x_2 \cup f_2(x_2)\} \wedge c \notin J(\lambda(x_1'), \lambda(x_2')) \wedge (x_1', x_2') \neq (x_1, x_2))\} \tag{4.5}$$

Initial state:

$$x_{||_0} = (x_{1_0}, x_{2_0}) \tag{4.6}$$

In equation 4.3, the cross product of the underlying NDAAO state spaces is built. $X_{||}$ consists of this cross product except of state combinations leading to a non-valid combined state output. Equation 4.4 assigns to each state of $X_{||}$ the combined partial I/O vectors of the underlying partial NDAAO states. In equation 4.5 the transition function of the cross product states is derived. Each cross product state (x_1, x_2) is connected to the state (x_1', x_2') if x_1' is either a successor of x_1 or x_1 itself and if x_2'

is either a successor of x_2 or x_2 itself. (x_1, x_2) is only connected to following states with valid outputs according to equation 4.2. This implements the synchronization of common I/Os. (x_1, x_2) is not connected to itself (no self-loop). In equation 4.6, the initial state is determined as the state with the underlying initial NDAAO states. If more than two NDAAO have to be synchronized, the cross product operation can be applied successively. Like in the case of the parallel composition from (Cassandras and Lafortune, 2006), it is possible to build the cross product of several NDAAO by applying the operation of definition 29 successively (associative law of the parallel composition):

$$\text{NDAAO}_1 || \text{NDAAO}_2 || \text{NDAAO}_3 = (\text{NDAAO}_1 || \text{NDAAO}_2) || \text{NDAAO}_3 \qquad (4.7)$$

An interesting property of the cross product language is given in the following theorem:

Theorem 6 (Completeness of the cross product of identified automata). We consider a closed-loop DES which is divided in N_{sys} subsystems. If the N_{sys} partial systems are identified such that $L_{Ident,sys_i}^{k+n} \supseteq L_{Orig,sys_i}^{k+n}$ $\forall 1 \leq i \leq N_{sys}$, it follows that $L_{Ident_{||}}^{k+n} \supseteq L_{Orig}^{k+n}$ for the language of the cross product.

This means that even if the complete system language cannot be considered as completely observed, it is possible to simulate it by the language of the cross product if the underlying partial languages can be considered as completely observed: According to theorem 5, $L_{Ident,sys_i}^{k+n} \supseteq L_{Orig,sys_i}^{k+n}$ holds if k was choosen such that $L_{Obs,sys_i}^{k+1} = L_{Orig,sys_i}^{k+1}$. For the proof, the definition of two functions is necessary. The first function replaces a substring consisting of equal symbols by a string of length 1, containing the considered symbol.

Definition 30 (Replace equal substrings function). Given a word w^n of length n. The function $w_{res}^{m \leq n} = \text{RemEqual}(w^n)$ replaces each substring $w_{sub}^2 \in w^n | w_{sub}^2 \langle 1 \rangle = w_{sub}^2 \langle 2 \rangle$ with[1] $w_{sub}^2 \langle 1 \rangle$. This operation is repeated until the resulting word $w_{res}^{m \leq n}$ no longer contains a w_{sub}^2 with $w_{sub}^2 \langle 1 \rangle = w_{sub}^2 \langle 2 \rangle$.

For the word $w^5 = ABBBC$, $w_{res}^{m \leq n} = \text{RemEqual}(w^n)$ delivers $w_{res}^3 = ABC$. The next function is the projection of an I/O vector to a subset of I/Os. It replaces each I/O in the vector by the don't care symbol '-' if it is not part of a given I/O set.

Definition 31 (I/O vector projection). Given an I/O vector $u(j)$ and a set of I/Os $IOSet$. The function $\text{IOProj}_{IOSet}(u(j))$ replaces the value of each $IO \in u | IO \notin IOSet$ by the don't care symbol '-'.

With $IOSet = y(sys_t)$ it is possible to derive the partial I/O vector of the t-th partial subsystem sys_t from the I/O vector of the global system $u_{sys_t} = \text{IOProj}_{y(sys_t)}(u)$ (with $y(sys_t)$ from definition 25).

---

[1]Like in chapter 3, $w^n \langle i \rangle$ addresses the i-th symbol (or letter) of the word w^n. This is not to be confused with the use of the $u[i]$ operation on single I/O vectors u to get the i-th I/O.

Proof of theorem 6. We proof the theorem for a closed-loop DES which is divided in two subsystems sys_1 and sys_2. The extension to more than two subsystems is straight forward. Each $w_{Orig}^{k+n} \in L_{Orig}^{k+n}$ can be seen as the result of two partial words $w_{Orig,sys_1}^{n_1 \le k+n} \in L_{Orig,sys_1}^{k+n}$, $w_{Orig,sys_2}^{n_2 \le k+n} \in L_{Orig,sys_2}^{k+n}$ of the two subsystems with

$$w_{Orig,sys_1}^{n_1 \le k+n} = \text{RemEqual}(\text{IOProj}_{y(sys_1)}(w_{Orig}^{k+n}))$$

$$w_{Orig,sys_2}^{n_2 \le k+n} = \text{RemEqual}(\text{IOProj}_{y(sys_2)}(w_{Orig}^{k+n}))$$

If an I/O vector is part of the original language, replacing all I/Os by '-' if they do not belong to the given subsystem leads to a partial I/O vector being part of the original subsystem language. From the precondition it follows that for the partial systems $L_{Ident,sys_1}^{k+n} \supseteq L_{Orig,sys_1}^{k+n}$ and $L_{Ident,sys_2}^{k+n} \supseteq L_{Orig,sys_2}^{k+n}$ holds. Hence, $w_{Orig,sys_1}^{n_1} \in L_{Ident,sys_1}^{k+n}$ and $w_{Orig,sys_2}^{n_2} \in L_{Ident,sys_2}^{k+n}$. From this it follows that for each $w_{Orig,sys_1}^{n_1}$ and $w_{Orig,sys_2}^{n_2}$ a state trajectory exists in the automaton that has been identified for the according subsystem, producing the original word. In equation 4.3, $X_{||}$ is constructed such that a cross product state exists for each combination of two partial automata states leading to a valid system output. Hence, for each combination of substrings $w_{Orig,sys_1}^1 \in w_{Orig,sys_1}^{n_1}$ and $w_{Orig,sys_2}^1 \in w_{Orig,sys_2}^{n_2}$ such that $\text{J}(w_{Orig,sys_1}^1, w_{Orig,sys_2}^1) \in w_{Orig}^{k+n}$, a cross product state exists with $\text{J}(w_{Orig,sys_1}^1, w_{Orig,sys_2}^1)$ as output. In equation 4.5, two cross product states $x_{||_1}$ with $\lambda_{||}(x_{||_1}) = w_{Orig}^{k+n}\langle j\rangle$ and $x_{||_2}$ with $\lambda_{||}(x_{||_2}) = w_{Orig}^{k+n}\langle j+1\rangle$ are connected if at least one of the underlying partial automaton states of $x_{||_1}$ has one of the underlying partial automaton states of $x_{||_2}$ as successor. Since for each $w_{Orig,sys_1}^{n_1}$ and $w_{Orig,sys_2}^{n_2}$ a state trajectory exists in the partial automaton of the according subsystem, the cross product contains a trajectory reproducing w_{Orig}^{k+n}. $\square$

From theorem 6 it follows that it is possible to identify an automata network which simulates the complete original system language although even L_{Obs}^1 (of the global system language) does not converge. The precondition is that the language of each of the considered subsystems converges as explained in section 4.1. Since the identified automata network is able to reproduce the complete original system language although $L_{Obs}^n \subset L_{Orig}^n$ ($\forall n \ge 1$), the number of false alerts during online diagnosis can be significantly reduced compared to the monolithic approach.

A disadvantage of the distributed approach is the increased exceeding language leading to a larger number of non-detectable faults. In the monolithic approach it was possible to guarantee that the exceeding language L_{Exc}^{k+1} is minimized. Although this property still holds for each of the identified partial automata, it can no longer be guaranteed for the language of the complete automata network. Even if each of the partial automata produces a subsystem word which has been observed before, the *combination* of fault-free subsystem words is not necessarily a word of the fault-free original system language. As an example to illustrate this phenomenon, the two partial automata in figure 4.6 are considered. They are identified for two subsystems considering the first two and the last two I/Os of the global I/O vector. The identification was carried out on the basis of the partial languages built with σ_1 in figure 4.6 according to definition 27. The identification parameter was chosen to $k = 1$.

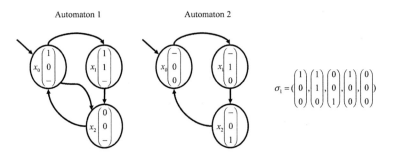

Figure 4.6: Example for exceeding behavior of an automata network

In the example, the second I/O is common to both of the partial automata. The observed sequence σ_1 which was the basis for the identification of the partial automata is depicted on the right of the automata network. The language of the automata network is defined by the cross product according to definition 29. The resulting automaton is depicted in figure 4.7.

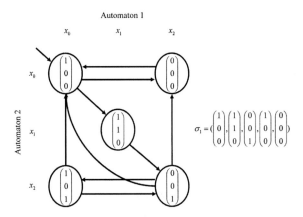

Figure 4.7: Cross product of the example

In the example it is supposed that under fault-free conditions, the underlying closed-loop DES can only exhibit the sequence σ_1. Any other sequence should thus not be part of the identified network. Suppose that a fault leads to the sequence

$$\sigma_F = (\begin{pmatrix} 1 \\ 0 \\ 0 \end{pmatrix}, \begin{pmatrix} 1 \\ 1 \\ 0 \end{pmatrix}, \begin{pmatrix} 0 \\ 0 \\ 1 \end{pmatrix}, \begin{pmatrix} 1 \\ 0 \\ 1 \end{pmatrix}, \begin{pmatrix} 0 \\ 0 \\ 1 \end{pmatrix})$$

Since the automata network (and thus the cross product) can perform a combined state evolution resulting in this sequence (x_0, x_1, x_2, x_0, x_2 for automaton 1 and x_0, x_1, x_2

for automaton 2), the fault cannot be detected. The sequence σ_F is part of the cross product language since it can be produced by a combined evolution of the partial automata.

The example shows that it is necessary to restrict the behavior of the automata network to prevent faulty behaviors from being part of the automata network language. In the next section it will be shown how the automata network can be restricted to producing only the observed global behavior and a well specified amount of unobserved behavior.

4.3 Restriction of an identified automata network

If the identified models are used for online fault detection, the current behavior of the system is compared with the model behavior. Only if both behaviors differ, a fault can be detected[2]. In order to reduce the number of non-detecable faults (faults with a behavior which can be reproduced by the model), the automata network language has to be systematically restricted. The restriction is performed based on a reduction of the cross product language. The first step is to determine the observed part of the cross product language. This part of the language is part of the original system language and should thus not be suppressed. To determine the observed part of the cross product language, a *transition observation function* is introduced:

Definition 32 (Transition observation function). The transition observation function $\Theta : X \times X \rightarrow \{true, false\}$ assigns to each pair (x_1, x_2) of NDAAO states *true* if a transition from x_1 to x_2 has been observed. If the transition has not been observed, the function assigns the value *false*. The decision is based on the set of observed sequences $\Sigma = \{\sigma_1, \ldots, \sigma_p\}$[3]. $\Theta(x_1, x_2)$ is undefined if $x_2 \notin f(x_1)$.

In the case of an identified monolithic NDAAO, the value of Θ is true for each transition in the automaton since in algorithm 1 states are only connected if their outputs have been observed successively. If the cross product of several identified partial automata is considered, transitions exist which do not correspond to an observed sequence. These transitions have their origin in the unsynchronized behavior of the underlying partial automata. The transition observation function can be determined with algorithm 3. The idea is to reproduce each observed sequence starting from the initial state of the cross product. If it is necessary to pass from a state $x_{\|1}$ to a state $x_{\|2}$ to reproduce the observed sequences, $\Theta(x_{\|1}, x_{\|2})$ is assigned to *true*. Otherwise, $\Theta(x_{\|1}, x_{\|2})$ is set to *false*.

At the beginning of algorithm 3, each transition is declared unobserved. For each observed sequence the following operations are performed: If the first I/O vector of a sequence is considered, the initial state of the cross product is taken as the current state $x_{\|}(j)$ (line 4 and 5). If the considered I/O vector is not the first one in the

---

[2]A more detailed description of the online diagnosis process using the identified models will be given in chapter 6
[3]see definition 19

Algorithm 3 Determination of the observed cross product

Require: Cross product $NDAAO_{||}$, observed sequences $\Sigma = \{\sigma_1, \ldots, \sigma_p\}$

1: $\forall (x_{||1}, x_{||2}) \in X_{||} \times X_{||} | x_{||2} \in f(x_{||1}) : \Theta(x_{||1}, x_{||2}) := false$

2: **for** each $\sigma_h \in \Sigma$ **do**

3:      **for** each $u_h(j) \in \sigma_h = (u_h(1), u_h(2), \ldots, u_h(l_h))$ **do**

4:          **if** $j = 1$ ($u_h(j)$ is the first vector in σ_h) **then**

5:              Current state $x_{||}(j) = x_{||0}$

6:          **else**

7:              $x_{||}(j) := x_{||} | (x_{||} \in f_{||}(x_{||}(j-1)) \wedge \lambda_{||}(x_{||}) = u_h(j))$

8:              $\Theta(x_{||}(j-1), x_{||}(j)) := true$

9:          **end if**

10:     **end for**

11: **end for**

---

sequence, the new current state $x_{||}(j)$ to reproduce the sequence is determined in line 7. The next state $x_{||}(j)$ is a following state of the former current state $x_{||}(j-1)$ and has the considered I/O vector as output. From step 2 of algorithm 1 (the monolithic identification algorithm) it follows that a state in an identified automaton does not have several following states with the same output. Hence, there is always only one state fulfilling the condition in line 7 of algorithm 3. In line 8, the transition between $x_{||}(j-1)$ and $x_{||}(j)$ is declared observed. If each sequence has been treated, the part of the cross product which corresponds to the observed system behavior can be obtained by applying Θ to the transitions in the automaton.

Figure 4.8 shows an example for the determination of the Θ function. It is the cross product from the last example. Based on the analysis of the observed sequence σ_1, Θ is determined with algorithm 3. The transitions with the solid lines have the Θ value *true* since they must be passed to reproduce the observed sequence. The transitions with the dashed lines are part of the cross product, but are not part of a state trajectory reproducing the observed sequence σ_1.

If the cross product performs a state trajectory containing unobserved transitions, this can lead to a word of the exceeding language $L_{Exc}^n = L_{Ident}^n \backslash L_{Orig}^n$. Since the combined behavior of the underlying partial automata resulting in the considered trajectory has not been observed before, it potentially represents a fault symptom. At the same time it is possible that a trajectory containing unobserved transitions represents a word of the original fault-free system language L_{Orig}^n which has not yet been seen. It should thus be accepted in order to avoid a false alert. A possible way to systematically parameterize the necessary trade-off between producing a certain amount of unknown behavior and avoiding a too important exceeding language is to *count* the number of unobserved transitions in a trajectory of the cross product. Hence, the restriction of the cross product language is possible by giving an upper limit of allowed unobserved transitions. This approach is based on the assumption that faults typically lead to a longer abnormal behavior. This assumption is common to most model-based diagnosis methods (e.g. the necessary deviant behavior to determine a diagnosis with the diagnoser approach of (Sampath et al., 1996), see chapter 2.3). The upper limit of

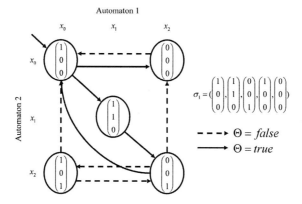

Figure 4.8: Example for an observed cross product

accepted unobserved behavior can be defined *manually* using an additional automaton of the NDAAO-type with a Θ function. The automaton can be seen as an additional parameter to influence the language of the identified model. The purpose of the automaton is to be run in a *synchronized* way with the identified system model containing observed and unobserved transitions and to serve as a counter:

Definition 33 (Tolerance specification). The tolerance specification is an

$$\text{NDAAO}_{Tol} = (X_{Tol}, \Omega_{Tol}, f_{Tol}, \lambda_{Tol}, x_{Tol_0})$$

with X_{Tol} finite set of states, $\Omega_{Tol} = \{\text{OK}, \text{Fault}, \text{-}\}$ output alphabet, $f_{Tol} : X_{Tol} \to 2^{X_{Tol}}$ non-deterministic transition function, $\lambda_{Tol} : X_{Tol} \to \Omega_{Tol}$ output function and x_{Tol_0} the initial state. In the output alphabet, OK represents a situation where the trajectory of the synchronized system model is accepted as fault-free. If the NDAAO$_{Tol}$ is in a state with the output Fault, the trajectory of the synchronized model is considered as a fault. '-' represents an undecided situation. For the tolerance specification, the transition observation function Θ must be defined manually for each transition.

The tolerance specification automaton has to be built manually. A possible implementation is given in figure 4.9 and will be explained after the following definition. To restrict the automaton network behavior, the tolerance specification automaton is run in a *synchronized* way with the cross product according to the following rule:

Definition 34 (Evolution rule for the cross product synchronized with a tolerance specification). The cross product can only change its current state $x_{||}(j)$ to $x_{||}(j+1)$, if there is a following state $x_{Tol}(j+1)$ of the current tolerance specification state $x_{Tol}(j)$ for which $\Theta(x_{||}(j), x_{||}(j+1)) = \Theta(x_{Tol}(j), x_{Tol}(j+1))$ holds. If such $x_{Tol}(j+1)$ and $x_{||}(j+1)$ exist, they become the next current states of the tolerance specification and the cross product respectively. If no such states exist, the cross product cannot proceed to another state.

In the tolerance specification a transition must be taken which has the same Θ-value like the transition in the cross product. If an appropriate transition does not exist, the cross product cannot enter the according state.

The language of the cross product generated synchronized with the tolerance specification contains two pieces of information. The first information is obtained by building the sequences of I/O vectors generated by state trajectories in the cross product. The second information is that for each of these sequences the output of the final state of the tolerance specification is known. Hence, it is known if a given sequence is considered as fault-free ('OK'), undecided ('-') or faulty ('Fault') by the tolerance specification.

With definition 34 it follows that a valid tolerance specification automaton that is able to restrict the cross product language must be conform to the following two rules:

1. Each state in the tolerance specification has at most two following states:

$$\forall x_{Tol} \in X_{Tol} : |f(x_{Tol})| \leq 2 \tag{4.8}$$

2. Two following states do not have the same Θ value. This makes sure that the state evolution rule in definition 34 can be correctly applied since the choice of the next current tolerance specification state is always unambiguous:

$$\left(\forall x_{Tol} \in X_{Tol} \wedge \forall x_{Tol_1}, x_{Tol_2} \in f(x_{Tol})\right) : \Theta(x_{Tol}, x_{Tol_1}) \neq \Theta(x_{Tol}, x_{Tol_2}) \tag{4.9}$$

Figure 4.9 shows an example of a tolerance specification automaton. When synchronized with a cross product according to definition 34, it tolerates passing one unobserved cross product transition (Θ-value is $false$). After the occurrence of a second unobserved transition, the tolerance specification automaton is in the state with the output 'Fault'. This state does not have any leaving transition. Hence, the evolution of the cross product is blocked according to definition 34.

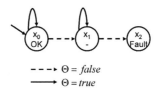

$$--- \blacktriangleright \ \Theta = false$$
$$\longrightarrow \ \Theta = true$$

Figure 4.9: Example for the tolerance specification automaton

As an example of the effect of the tolerance specification we consider again the faulty sequence

$$\sigma_F = (\begin{pmatrix} 1 \\ 0 \\ 0 \end{pmatrix}, \begin{pmatrix} 1 \\ 1 \\ 0 \end{pmatrix}, \begin{pmatrix} 0 \\ 0 \\ 1 \end{pmatrix}, \begin{pmatrix} 1 \\ 0 \\ 1 \end{pmatrix}, \begin{pmatrix} 0 \\ 0 \\ 1 \end{pmatrix})$$

Reproducing σ_F, the cross product in figure 4.8 passes two unobserved transitions: after the trajectory $x_0 x_0 \rightarrow x_1 x_1 \rightarrow x_2 x_2$ with exclusively observed transitions, the trajectory $x_2 x_2 \rightarrow x_0 x_2 \rightarrow x_2 x_2$ to reproduce the last three I/O vectors in σ_F contains two unobserved transitions. If the tolerance specification is run in a synchronized way, the two unobserved transitions lead it to the state with the output 'Fault' which declares that the considered sequence is not part of the fault-free behavior. With the tolerance specification it is thus possible to define the amount of not yet known global system behavior which is considered as a fault symptom although each partial behavior is known. With the tolerance specification of figure 4.9, the following sequence σ_2 would not be considered as a fault since it leads to only one unobserved transition in the trajectory $x_0x_0 \to x_0x_2 \to x_0x_0$ in figure 4.8.

$$\sigma_2 = (\begin{pmatrix} 1 \\ 0 \\ 0 \end{pmatrix}, \begin{pmatrix} 0 \\ 0 \\ 0 \end{pmatrix}, \begin{pmatrix} 1 \\ 0 \\ 0 \end{pmatrix})$$

The tolerance specification allows to balance the number of false alerts and the sensitiveness to faults. In the example it was decided that sequences leading to only one yet unknown transition in the cross product are probably part of the fault-free behavior whereas a higher number of unobserved transitions is not acceptable as fault-free behavior. Choosing an appropriate structure of the tolerance specification automaton is a degree of freedom to adjust the model to a given system and the available observed system language. In section 4.4 it will be shown how an analysis of the observed system language can help to determine an appropriate tolerance specification. Nevertheless, the choice of the tolerance specification will always depend on the preferences of the user.

The approach of constructing the cross product (and to derive the observed part of it) has an important disadvantage if it is to be applied to large systems: building the cross product of several partial automata can lead to the state space explosion problem. According to (Bérard et al., 2001), the state space resulting from composing several automata with state spaces X_1, X_2, ..., X_n can reach up to $|X_1| \times |X_2| \times \cdots \times |X_n|$ states. For practical implementation, this poses two problems. The first problem is that the procedure to obtain the cross product from definition 29 can take very long. The second problem is that for diagnosis purposes, the resulting cross product must usually be handled online. If the resulting automaton is too large, it can be difficult to apply diagnosis procedures to the model in real time. Since the described approach is based on the distinction between the *observed* and the *unobserved* part of the cross product, it is not necessary to build and maintain the complete cross product. In the remainder of this section it is shown how it is possible to only construct the observed part of the cross product explicitly and to derive the unobserved part implicitly when it is necessary to produce a given part of the unobserved cross product language. The algorithms are an improved version of the procedures given in (Roth et al., 2009a).

The first step of the proposed approach is to directly build the *observed* part of the cross product based on the available observed sequences $\Sigma = \{\sigma_1, \ldots, \sigma_p\}$ and the identified partial automata. The *observed cross product* resulting from algorithm 4 is an NDAAO$_{Obs\|}$ with the output alphabet $\Omega_{Obs\|} = X_1 \times X_2 \times \cdots \times X_n$: The output alphabet (and thus the state outputs) consist of state combinations from the n partial automata from which the observed cross product is built. With this definition of the state outputs it is possible to determine for each NDAAO$_{Obs\|}$-state the underlying state combination of the partial automata by applying the state output function $\lambda_{Obs||}$. With the join-function from definition 28, it is possible to get the I/O vector corresponding to the combination of partial automaton states represented by a given observed cross product state.

In contrast to the construction of the cross product $\text{NDAAO}_{||}$, in algorithm 4 not each combination of partial NDAAO states leads to the creation of a new state in $\text{NDAAO}_{Obs||}$. Only combinations which are necessary to reproduce the already observed behavior (I/O vector sequences) are considered. For each of these combinations a state in the observed cross product is built. By successively analyzing the observed sequences and 'playing' the partial automata, the observed cross product states are connected in the necessary order. The resulting automaton only contains the observed substructure of the cross product from definition 29.

Algorithm 4 Direct construction of the observed cross product

---

Require: Set of identified partial automata $\{\text{NDAAO}_1, \ldots, \text{NDAAO}_n\}$ and observed sequences $\Sigma = \{\sigma_1, \ldots, \sigma_p\}$ according to definition 19

1: Initialize $\text{NDAAO}_{Obs||}$ with $X_{Obs||} = \{\}$

2: Create initial state $x_{Obs||0} | \lambda_{Obs||}(x_{Obs||0}) := \{x_{1\_0}, \ldots, x_{n\_0}\} \wedge f_{Obs||}(x_{Obs||0}) := \{\}$

3: **for** each $\sigma_h \in \Sigma$ **do**

4:     **for** each $u_h(j) \in \sigma_h = (u_h(1), u_h(2), \ldots, u_h(l_h))$ **do**

5:         **if** $j = 1$ ($u_h(j)$ is the first vector in σ_h) **then**

6:             Initialize current states $x_1(1) := x_{1\_0}, \ldots, x_n(1) := x_{n\_0}$ in the partial automata

7:             Initialize current state in the observed cross product: $x_{Obs||}(1) := x_{Obs||0}$

8:         **else**

9:             **for** each $\text{NDAAO}_i \in \{\text{NDAAO}_1, \ldots, \text{NDAAO}_n\}$ **do**

10:                 $x_i(j) := x_i | (x_i \in \{x_i(j-1) \cup f_i(x_i(j-1)) \wedge c \notin J(\lambda(x_i), u_h(j)))$

11:             **end for**

12:             **if** $\exists x_{Obs||} \in X_{Obs||} | \lambda_{Obs||}(x_{Obs||}) = \{x_1(j), \ldots, x_n(j)\}$ **then**

13:                 $f_{Obs||}(x_{Obs||}(j-1)) := \{f_{Obs||}(x_{Obs||}(j-1)) \cup x_{Obs||}\}$

14:             **else**

15:                 Create new $x_{Obs||} | (\lambda_{Obs||}(x_{Obs||}) = \{x_1(j), \ldots, x_n(j)\}, f_{Obs||}(x_{Obs||}) = \{\})$

16:                 $f_{Obs||}(x_{Obs||}(j-1)) := \{f_{Obs||}(x_{Obs||}(j-1)) \cup x_{Obs||}\}$

17:             **end if**

18:             $x_{Obs||}(j) = x_{Obs||} | \lambda_{Obs||}(x_{Obs||}) = \{x_1(j), \ldots, x_n(j)\}$ (either found in line 12 or created in line 15)

19:         **end if**

20:     **end for**

21: **end for**

In line 1 of algorithm 4, the observed cross product $\text{NDAAO}_{Obs||}$ is initialized with an empty state space. In the next step, the initial observed cross product state is constructed. Its output function is set to the set of initial states of the underlying partial automata. Starting in line 3, the observed sequences are analyzed. At the beginning of each sequence, the partial automata are initialized by setting their current state to the initial partial automaton state (e.g. for the first partial automaton $NDAAO_1$: $x_1(1) = x_{1\_0}$). The observed cross product is also initialized by setting its current state $x_{Obs||}(1)$ to its initial state $x_{Obs||_0}$. This is always done when the first I/O vector of a sequence is treated (line 4 to 7). If the considered I/O vector is not the first one in the sequence, the algorithm proceeds in line 9. From line 9 to line 11, the current state of each partial NDAAO is determined. A partial NDAAO state becomes the new current state $x_i(j)$ if it is either the former current state $x_i(j-1)$ or one of its successor states ($x_i \in \{x_i(j-1) \cup f_i(x_i(j-1))\}$). Additionally, it must have an output that corresponds to the considered I/O vector $u_h(j)$. Since in the partial automaton $NDAAO_i$ not all I/Os are considered (I/Os not belonging to the according subsystem are replaced by '-' symbols in definition 26), the state outputs and the I/O vector of the sequence are compared with the join-function[4]. If the join-function returns 'c' (contradiction) for the comparison of at least one I/O value, the state output and the I/O vector differ in at least one I/O considered in the partial $NDAAO_i$. Hence, if $c \notin J(\lambda(x_i), u_h(j))$, the state output of the partial automaton can reproduce its part of the current I/O vector.

The states from the set $x_i \in \{x_i(j-1) \cup f_i(x_i(j-1))\}$ have different state outputs due to step 2 of algorithm 1. Hence, for each $NDAAO_i$ there is always a unique state fulfilling the condition in line 10. In line 12 it is checked if the observed cross product already contains a state with the newly determined combination of current partial automaton states as output. If so, the former current state of the observed cross product $x_{Obs||}(j-1)$ gets this state as a following state. If not, a new state is created. Its output function is set to the combination of current partial automaton states. In line 16, the newly created state is added to the set of following states of the former current observed cross product state. Finally, in line 18, the current state is set to the newly determined state and the next I/O vector is considered.

An example for the result of algorithm 4 is given in figure 4.10. On top of the figure, the observed sequence σ_1 and two partial automata identified based on this sequence are shown. In the first step of algorithm 4 the initial state of the observed cross product is built. Its output is set to the initial states of the underlying partial automata $x_0 x_0$. When sequence σ_1 is treated, the partial automata perform the combined state trajectory $x_0 x_0$, $x_1 x_1$, $x_2 x_2$, $x_0 x_0$ and $x_2 x_0$. For each of these combinations a state in the observed cross product is built. The observed cross product states are connected according to the combined state trajectory of the underlying partial automata. The result is the automaton in the bottom of figure 4.10.

In the observed cross product, each transition has been observed since it has been created upon an analysis of the observed sequences. Hence, the transition observation function Θ can be set *true* for the complete automaton structure. In order to apply the tolerance specification to define a certain amount of accepted unobserved behavior, it is necessary to add unobserved transitions to the observed cross product. This is done in the next algorithm. The result is the permissive observed cross product:

Definition 35 (Permissive observed cross product (POCP)). The permissive observed cross product is a 6-tuple $POCP = (X_{POCP}, \Omega_{POCP}, f_{POCP}, \lambda_{POCP}, x_{POCP_0}, \Theta_{POCP})$

---

[4]See definition 28.

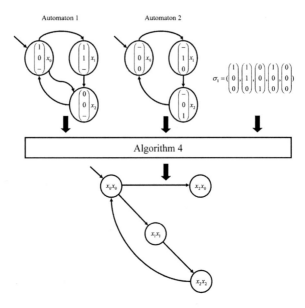

Figure 4.10: Example for algorithm 4

with X_{POCP} the state space, Ω_{POCP} the output alphabet, $f_{POCP} : X_{POCP} \rightarrow 2^{X_{POCP}}$ the transition function, $\lambda_{POCP} : X_{POCP} \rightarrow \Omega_{POCP}$ the output function and x_{POCP_0} the initial state of the automaton. $\Theta_{POCP} : X_{POCP} \times X_{POCP} \rightarrow \{true, false\}$ is the transition observation function.

Based on the observed cross product resulting from algorithm 4, the POCP is constructed with algorithm 5. At the beginning of the algorithm, the output alphabet of the observed cross product is assigned to the POCP output alphabet. Additionally, the symbol ε is added to Ω_{POCP} to represent each non-observed combination of partial automata. In the next steps the basic state space, the basic transition function and the output function are constructed by assigning $X_{Obs\|}$ to X_{POCP}, $f_{Obs\|}$ to f_{POCP} and $\lambda_{Obs\|}$ to λ_{POCP}. Then, the transition observation function is defined. Since until this moment the POCP only contains the observed cross product, Θ_{POCP} is assigned to $true$ for each pair of states that is connected by a transition. In step 5 a new state x_ε with the empty letter ε as output is created and in step 6 added to the state space. In the next step, x_ε gets the complete state space as following states. Hence, there is a transition from x_ε to each other state in X_{POCP} as well as a self-loop to x_ε itself. The state x_ε is added to the POCP state space to represent the unobserved part of the cross product. In the next step, the transition function of the states in X_{POCP} except of the formerly created x_ε is enlarged. Each state is connected to each other state. There are no self-loops added. In the last part of the algorithm the transition observation function is completely defined. For each transition in the POCP with an undefined Θ_{POCP} function, Θ_{POCP} is set to $false$ since the transition was not part of the observed cross product.

Algorithm 5 Construction of the permissive observed cross product

Require: Observed cross product NDAAO$_{Obs\|}$
1: Output alphabet: $\Omega_{POCP} := \Omega_{Obs\|} \cup \varepsilon$ with ε being the empty output
2: Basic state space: $X_{POCP} := X_{Obs\|}$, $x_{POCP_0} = x_{Obs\|_0}$
3: Basic transition function / output function: $f_{POCP} := f_{Obs\|}$, $\lambda_{POCP} := \lambda_{Obs\|}$
4: Transition observation function: $\forall x \in X_{POCP} : \Theta_{POCP}(x, f_{POCP}(x)) := true$
5: Create state with empty output: $x_\varepsilon : \lambda_{POCP}(x_\varepsilon) := \varepsilon$, $f_{POCP}(x_\varepsilon) = \{\}$
6: Enlarged state space: $X_{POCP} := X_{POCP} \cup x_\varepsilon$
7: Following states of x_ε: $f_{POCP}(x_\varepsilon) := X_{POCP}$
8: Enlarged transition function: $\forall(x \neq x_\varepsilon) \in X_{POCP} : f_{POCP}(x) := X_{POCP}\backslash x$
9: Complete definition of the transition observation function:
10: **for** each $x \in X_{POCP}$ **do**
11:    **for** each $x' \in f_{POCP}(x)$ **do**
12:       **if** $\Theta_{POCP}(x, x')$ not defined **then**
13:          $\Theta_{POCP}(x, x') := false$
14:       **end if**
15:    **end for**
16: **end for**

Figure 4.11 shows the POCP created by algorithm 5 based on the observed cross product from figure 4.10. The black state represents the state x_ε added during the algorithm. It can be seen that each state is reachable by each other state by a transition. If this transition was already existent after the construction of the observed cross product (algorithm 4), the according Θ_{POCP} value is *true*. If a transition has been added during algorithm 5, the according Θ_{POCP} value is *false*.

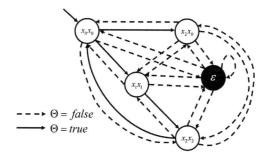

Figure 4.11: Example for a POCP

The POCP can now be run in parallel with its underlying network of partial NDAAO and a given tolerance specification to restrict the network behavior. The principle is shown in figure 4.12. The automata network can perform any combined state trajectory leading to a valid model output[5]. The POCP evolves following the automata network. It tries to perform a state trajectory reproducing the current trajectory in the automata network. For each new state combination produced by the network, the POCP tries to find a state x_{POCP} that has the current combination of partial NDAAO states as output. If the network trajectory has state combinations which are not represented by a POCP state, the POCP takes the state with the empty symbol ε as output instead. By analyzing the transitions passed in the POCP (observed or unobserved ones), the tolerance specification evolves such that it passes the same type of transitions. For each combined state trajectory of the automata network it can thus be decided if it is accepted as fault-free or if it is considered as a fault from the output of the reached state in the tolerance specification. In definition 36 the combined evolution of partial automata, POCP and tolerance specification is formally described.

Definition 36 (Evolution rule for the POCP in parallel with a tolerance specification and a network of partial NDAAO). Let the n partial NDAAO be in the combined state $\{x_1(j), \ldots, x_n(j)\}$ with $c \notin J(\lambda_1(x_1(j)), \ldots, \lambda_n(x_n(j)))$. If a POCP state x_{POCP} with $\lambda_{POCP}(x_{POCP}) = \{x_1(j), \ldots, x_n(j)\}$ exists, this state is the current POCP state $x_{POCP}(j)$. If such a state does not exist, the state with the empty output ε is the current POCP state: $x_{POCP}(j) = x | \lambda_{POCP}(x) = \varepsilon$. The n partial NDAAO can move from their current state combination $\{x_1(j), \ldots, x_n(j)\}$ to another state combination $\{x_1(j+1), \ldots, x_n(j+1)\}$ with $c \notin J(\lambda_1(x_1(j+1)), \ldots, \lambda_n(x_n(j+1)))$ if $\forall 1 \leq a \leq n$ either $x_a(j) = x_a(j+1)$ or $x_a(j+1) \in f_a(x_a(j))$ holds[6]. The POCP evolves according to the following rule:

$$\textbf{If} \quad (\exists x'_{POCP} \in X_{POCP}) | (\lambda_{POCP}(x'_{POCP}) = \{x_1(j+1), \ldots, x_n(j+1)\}) :$$
$$x_{POCP}(j+1) := x'_{POCP}$$
$$\textbf{else} \quad x_{POCP}(j+1) := x_{POCP} | \lambda_{POCP}(x_{POCP}) = \varepsilon$$

The tolerance specification moves from its current state $x_{Tol}(j)$ to a following state $x_{Tol}(j+1)$ with

$$x_{Tol}(j+1) \in f_{Tol}(x_{Tol}(j)) \wedge \Theta_{POCP}(x_{POCP}(j), x_{POCP}(j+1)) = \Theta_{Tol}(x_{Tol}(j), x_{Tol}(j+1))$$

If such a state does not exist, the model consisting of POCP, partial automata and tolerance specification cannot proceed to another state.

With the evolution rule of definition 36, the language of the automata network restricted by the POCP and the tolerance specification has two elements. The I/O vector sequences building the identified language are determined by applying the join-function to the partial state outputs from the set of current partial NDAAO states. If an I/O vector sequence is generated by the restricted automata network, the combined state trajectory leads to a trajectory in the POCP consisting of observed and unobserved transitions. Since the tolerance specification evolves in a synchronized way by

---

[5] A valid model output is determined by applying the join-function to the partial state outputs. If c is not part of the join-function, the combined output is valid

[6] Each partial NDAAO must either stay in its current state or it takes the next state from the following states of the current state

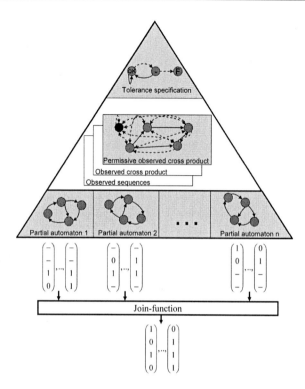

Figure 4.12: Principle of the restricted automata network with the POCP

the transition observation function Θ_{Tol}, the combined state trajectory leading to the given I/O vector sequence results in a certain state in the tolerance specification. From the output of this state, it can be decided if the considered I/O vector sequence is considered as fault-free (state output is 'OK'), not faulty in the sense of undecided ('-') of faulty ('Fault').

With the tolerance specification it is thus possible to explicitly define the amount of yet unseen automata network behavior which is considered as fault-free. Hence, it is possible to deal with a certain amount of new system behavior without directly detecting a fault which avoids a large number of false alerts. The presented approach allows accepting some behavior which is *similar* but not *equal* to the already known behavior. The tolerance specification defines the *degree of similarity* to some known fault-free behavior which makes a given behavior not be considered as faulty. This concept is well known in the field of diagnosis of continuous systems. Models of continuous systems do not perfectly reproduce signal values. It is thus necessary to define tolerance limits to decide if the difference between the measured and the modeled signal value is a fault symptom. The tolerance specification can be understood as the definition of these limits for the identified automata network.

The fact that each accepted behavior must consist of the combination of fault-free subsystem behaviors (defined by the partial automata) is an important advantage: it avoids considering *any arbitrary* behavior as similar to the fault-free behavior. Only behaviors which consist of correct partial behaviors can be tolerated. In section 6.2, a fault detection policy will be presented which makes use of this advantage: Observed behavior which cannot be explained by a combination of fault-free subsystem behavior defined by the partial automata will be *directly* considered as a fault. If the observed behavior consists of unknown combined subsystem behavior, the tolerance specification will be evaluated.

Remark 3 (Subsystem granularity). The advantage of not considering *any unknown behavior* as similar to the fault-free behavior becomes less important if the number of subsystems is increased which usually leads to smaller subsystems. If each I/O gets its own subsystem (as an extreme example for system partitioning), the resulting partial automata will consist of two states: one state represents $IO_i = 0$ and one state represents $IO_i = 1$. It is obvious that this makes fault detection on the subsystem level impossible: The subsystems do not contain any other I/O which would make it possible to determine if IO_i changed its value to early or to late. The lower the number of subsystems is and the larger the subsystems in terms of I/Os belonging to them are, the more likely it is that a faulty behavior is already refused on the subsystem level. Hence, it can be concluded that *the number of subsystems should be kept as small as possible.*

4.4 Distributed identification of the case study system

The distributed identification approach has been applied to the case study system from chapter 3.5. The subsystems have been chosen as shown in figure 4.13 based on apriori knowledge. In each subsystem the controller I/Os belonging to one of the machine tools with the according conveyor have been grouped. Each subsystem consists of components with a direct physical relation like the single machine tools and the conveyor in front of them. The position sensor between the first two machine tools is assigned to both subsystems since it belongs to both of the conveyors. The same consideration holds for the sensor between the second and the third machine tool. A detailed description of how to choose appropriate subsystems based on a priori knowledge or by analyzing the observed system language will be given in chapter 5.

For each subsystem depicted in figure 4.13 the observed language is depicted in figure 4.14. It can be seen that each subsystem language directly converges to a stable level. The reason for this fast convergence is the purely sequential behavior of each machine tool. Each subsystem leads to exactly the same sequence of partial I/O vectors in each observed system evolution. In section 4.1 it was explained that the aim of the distributed identification approach is to determine a model which is able to reproduce the complete global system behavior although the observed global language has not yet converged to a stable level (see figure 4.1 on page 63 for the evolution of the global

First machine tool   Second machine tool   Third machine tool

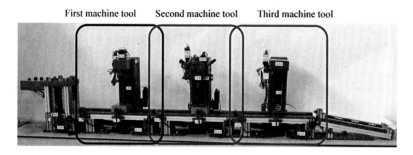

Figure 4.13: Laboratory system with subsystems

system language). With the result of theorem 6 it is thus possible to identify partial automata for each of the subsystems which build an automata network simulating the complete fault-free system behavior. The necessary condition of theorem 6 is fulfilled since each subsystem language converges.

In order to validate the presented approach, the identification data base is divided in two parts. The first 70 observed system evolutions determine the new identification database. The remaining 30 observed evolutions represent a part of the not yet seen original system language. In figure 4.1 (page 63) it can be seen that during these last evolutions the observed language grows even for small values of n. It will be tested if the identified automata network in conjunction with the POCP and an appropriately chosen tolerance specification is able to reproduce the language represented by the last 30 observed evolutions although the model has only been identified on the basis of the first 70 observed evolutions.

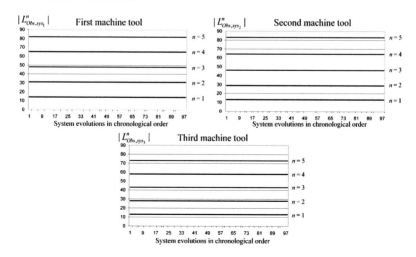

Figure 4.14: Observed language of the subsystems

For the distributed approach it is necessary to identify a partial automaton for each subsystem. Figure 4.14 shows that following the considerations of subsection 3.4.3, high values for the identification parameter k are possible since even for high values of k, L_{Obs}^{k+1} can be considered as completely observed in each subsystem. In order to determine an appropriate value for the parameter k, figure 4.15 shows the transition gap-function from definition 23 for each subsystem. It can be seen that in each case increasing k to $k = 2$ leads to a significantly decreased value in the gap function. Following the considerations in chapter 3.4.3, $k = 2$ is a reasonable choice for each partial automaton of the subsystems.

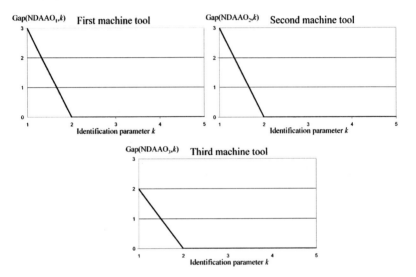

Figure 4.15: Transition gap for the subsystems of the case study

The partial automata are thus identified based on the first 70 system evolutions with parameter $k = 2$. Since in none of the partial languages of length $k + 1$ new words are observed in the last 30 evolutions, theorem 6 assures that the automata network is able to reproduce the complete global original system language. Consequently, it could be validated successfully that the (unrestricted) cross product of the automata network indeed contains each new word having been observed during the last 30 evolutions.

After identification of the partial NDAAO representing the subsystems, the next step is to construct the upper structure in the pyramid-schema in figure 4.12 to implement the automata network restriction. In section 4.3, two possible ways to determine this upper structure have been shown. The first one consists of the explicit construction of the cross product from which the observed part is derived. Constructing the cross product according to the procedure given in definition 29 leads to an automaton with 2565 states. The construction takes 5.7 seconds[7]. Deriving the observed cross product

---

[7] 1.8 GHz, 2 GB RAM

based on the first 70 observed evolutions according to algorithm 3 takes 0.2 seconds.

The alternative approach presented in the former section consists in directly constructing the observed cross product from the observed evolutions and the identified partial automata according to algorithm 4. Applying the algorithm using the identified automata network and the first 70 evolutions takes 1.5 seconds and leads to an automaton with 229 states. Based on this automaton, algorithm 5 constructs the POCP within a few milliseconds.

It can be seen that the second approach delivers the necessary models faster than the first one (1.5 in contrast to 5.7 seconds). More important than the gain in calculation time is the reduced size of the automaton. In the first approach it is necessary to manage an automaton with 2565 states during online diagnosis whereas in the second approach only the partial automata with 17, 15 and 15 states as well as the POCP with 230 states (229 plus the 'joker'-state added during algorithm 5) have to be handled online to generate the system language.

The last step of the distributed identification approach is the determination of the tolerance specification. It must be determined how many unobserved transitions in the POCP are tolerated without considering a generated word as a fault. Increasing the tolerance helps avoiding false alerts but comes at risk of erroneously considering faulty words as fault-free. Finding the right balance depends on the preferences of the user. In many cases the main focus may be to avoid false alerts as far as possible but to choose the tolerance specification as restrictive as possible. In order to assess the necessary number of tolerated unobserved POCP transitions, figure 4.16 is helpful. The figure considers the evolution of the observed cross product constructed during algorithm 4 (direct construction of the observed cross product). On the abscissa the observed system evolutions[8] are given. On the ordinate the number of new observed transitions created by analyzing each new sequence in algorithm 4 is depicted. It can be seen that during the first evolutions many new transitions are created since the automata network performs many new combined state trajectories. During the last ten evolutions, between zero and three new transitions are added to the observed cross product (see the zoom in figure 4.16). This implies that in future system evolutions only a few new combined state trajectories in the automata network are necessary to reproduce the observation. A possible choice for the tolerance specification is thus to accept for example one new combined movement of the automata network leading to one unobserved transition in the POCP. Evolutions like the 62th one would thus not lead to fault detection. An appropriate translation of this into a tolerance specification automaton is given in the right part of figure 4.16.

In order to evaluate the efficiency of the choice of the tolerance specification, the last 30 observed system evolutions have been reproduced with the automata network and the POCP according to definition 36. Before an evolution is reproduced, each partial automaton, the POCP and the tolerance specification are reset to their initial state. Reproducing 17 of 30 evolutions was possible by ending in the OK-state of the tolerance specification. The reproduction of seven evolutions ended in the undecided state. One

---

[8]Each evolution is an I/O vector sequence in $\Sigma = \{\sigma_1, \ldots, \sigma_p\}$

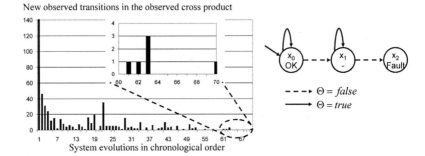

Figure 4.16: New observed transitions per evolution in algorithm 4 and the derived tolerance specification

unobserved transition was necessary to reproduce the according observations. In six evolutions, the reproduction ended in the fault state of the tolerance specification. Note that in each of the evolutions the partial automata were able to reproduce their partial language. Only the combined network trajectory was not accepted by the tolerance specification. This test shows that the number of false alerts (which occurred in 6 from 30 evolutions) is acceptable.

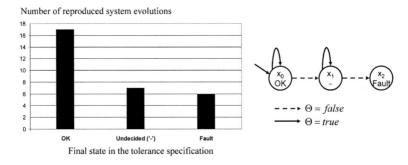

Figure 4.17: Efficiency of the tolerance specification tolerating one unobserved network trajectory

If the number of tolerated unknown network trajectories is increased to two, the number of false alerts can be decreased. Figure 4.18 shows the resulting number of fault detections. It can be seen that the number of evolutions ending in the 'Fault'-state of the tolerance specification is decreased to four. To compare the efficiency of the distributed and the monolithic model in terms of avoiding false alerts, a monolithic NDAAO has been identified on the basis of the first 70 system evolutions. The identification tuning parameter was set to $k = 1$ (although figure 4.1 shows that the according observed language does not converge). With the monolithic automaton it was not possible to reproduce 13 of 30 remaining system evolutions which corresponds to 13 false alerts.

This shows that using the distributed approach it is possible to decrease this number significantly (in this case from 13 to 4).

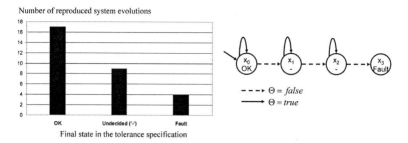

Figure 4.18: Efficiency of the tolerance specification tolerating two unobserved network trajectories

The fact that the combined evolution according to definition 36 leads to false alerts shows that adding the upper structure in the pyramid (figure 4.12) increases the sensitiveness to faults. Global system behavior with an important difference to already observed behavior is considered as a fault although it consists of valid subsystem trajectories. Faults leading to this kind of behavior can be detected with the pyramid structure but not with partial automata only. Increasing the number of tolerated unknown network trajectories like in the case of figure 4.18 decreases the number of false alerts, but also weakens the fault detection capability since a larger part of the cross product language can be generated in this scenario.

5 Partitioning of Discrete Event Sub-Systems

5.1 Characteristics of appropriate subsystems for diagnosis purposes

In the former chapter it has been shown that the efficiency of the identification approach can be significantly increased by dividing a given system in subsystems. Dividing a system in subsystems is also a very common technique if models are built manually. In section 2.3.2, it has been shown that the model-building process of the diagnoser approach relies on the construction of component models (e.g. for the valve in figure 2.6) which can be interpreted as subsystems. The question of how to determine the subsystems in the diagnoser approach refers to choosing the correct events for the component models. In (Philippot et al., 2007) it is proposed to choose components such that they contain one actuator (and the according actuator events) and its associated sensors (and the according sensor events). The idea is to represent causal relations in the single component models. For diagnosis purposes this kind of representation is advantageous since many faults lead to acausal system behavior which cannot be reproduced by a model representing fault-free causal relations. In section 4.2 it has been explained that choosing appropriate subsystems for closed-loop DES refers to selecting appropriate controller I/Os for each subsystem. Controller outputs are usually associated with actuators whereas controller inputs are associated with sensors. The following example shows how the representation of causal relations in a subsystem model facilitates diagnosis.

In figure 5.1 a double acting cylinder is shown. The cylinder has the two position sensors Pos1 and Pos2 which are connected to the first two inputs of the controller (they are the first two I/Os in the controller I/O vector). The actuator is connected to the two controller outputs Extend and Retract (the last two I/Os in the controller I/O vector). In the example it is supposed that the position sensors deliver 1 if they detect something and that actuators start an action upon receiving 1 from the controller. In the figure, two parts of partial automata representing possible subsystems[1] are shown. In the first subsystem, the two position sensors are represented. The leftmost partial automaton represents the behavior of the two sensors when the cylinder is extended: Pos1 changes its value to 0 and upon arrival of the cylinder at the extended position Pos2 changes its value to 1. In the example we consider the following fault: position

---

[1]Following the considerations of (Philippot et al., 2007), the cylinder would not be divided into subsystems since its actuator has an influence on both position sensors. In the example we nevertheless divide it in two subsystems to show the impact on diagnosis

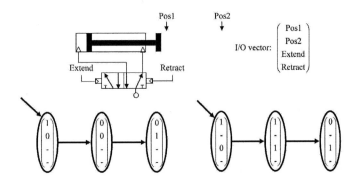

Figure 5.1: Double acting cylinder with paritial NDAAOs

sensor Pos1 is damaged such that it changes its value from 1 to 0 without the cylinder being extended. This produces the following faulty sequence:

$$\sigma_F = (\begin{pmatrix} 1 \\ 0 \\ 0 \\ 0 \end{pmatrix}, \begin{pmatrix} 0 \\ 0 \\ 0 \\ 0 \end{pmatrix})$$

If only the I/Os of the first subsystem are considered, the sequence is:

$$\sigma_{F_{Subsystem1}} = (\begin{pmatrix} 1 \\ 0 \\ - \\ - \end{pmatrix}, \begin{pmatrix} 0 \\ 0 \\ - \\ - \end{pmatrix})$$

This sequence can be reproduced by the leftmost partial automaton by moving from the first state to the second state. Hence, the fault cannot be detected. This situation changes if the second subsystem consisting of Pos1 and Extend is considered. A part of the system behavior is modeled by the rightmost partial automaton in figure 5.1. If only the I/Os Pos1 and Extend are considered, σ_F results in

$$\sigma_{F_{Subsystem2}} = (\begin{pmatrix} 1 \\ - \\ 0 \\ - \end{pmatrix}, \begin{pmatrix} 0 \\ - \\ 0 \\ - \end{pmatrix})$$

This sequence cannot be reproduced by the rightmost automaton. Hence, the fault can be detected. The second subsystem is more appropriate for detecting the fault at Pos1 since it covers the causal relation between the controller output Extend and the sensor Pos1. Since Pos1 should only change its value if Extend already changed its value *before*, a change in value at Pos1 *without* setting Extend before can be detected. The causal relation between Pos1 and Extend implies a required event order during fault-free behavior. This precise order is not captured in the first but in the second subsystem.

The reason for dividing a given system into subsystems in the distributed identification approach is to reduce the number of false alerts during online fault diagnosis. In chapter 4 it has been explained that an important requirement for the subsystems is to have a converging observed subsystem language. As shown in theorem 6, this leads to a minimized number of false alerts during online diagnosis with the identified model since the distributed model is able to produce words of the original global system language which have not been observed before. If the subsystems are build such that they exclusively contain I/Os with causal relations, this often leads to sequential subsystem behavior. Most often, an actuator only influences a limited number of sensors in a well defined order (like in the position sensors of the double acting cylinder). As lined out in section 4.1, sequential behavior leads to a manageable size of the subsystem language resulting in fast convergence. In the next section, some guidelines of how to choose appropriate subsystems based on a priori knowledge are given. Since this knowledge is often not available, section 5.3 presents an approach to automatically divide a given system into subsystems with weak internal concurrency by analyzing the observed system language. It is shown how (possibly partial) knowledge about causal relations in the considered system can be integrated to improve the result of the automated partitioning. In the last part of the chapter, the presented partitioning approach is applied to the case study system.

5.2 Partitioning based on a priori expert knowledge

In chapter 4 it has been explained that subsystems must be chosen such that the original subsystem language can be completely observed within a small number of system evolutions which leads to a converging observed subsystem language. The Petri net example in section 4.1 showed that increasing the concurrency in a system increases the size of the original system language. The larger the original system language is, the more system evolution it takes to observe a significant part of it. Hence, the observed language of subsystems with a low degree of internal concurrency typically converges faster than the language of subsystems with many concurrent parts. The challenge when dividing a system manually is to determine subsystems such that they lead to weakly concurrent internal subsystem behavior.

A possible way to get the necessary information for manual partitioning of a closed-loop DES is to analyze its specification. A widely used specification language for closed-loop DES in industry is Grafcet (Graphe Fonctionnel de Commandes, Etapes, Transitions) of SFC (Sequential Function Chart)(IEC, 2002). With this tool it is possible to explicitly code that two (or more) sequences of actions can be executed simultaneously. Figure 5.2 shows an example for a Grafcet with two parallel sequences of actions. It can be seen that after step 10, two parallel branches exist. Hence, the sequences in the two branches are to be executed in parallel. The sequences in two parallel branches are usually related to disjunct sets of controller I/Os. Hence, it is possible to assign the controller I/Os from one branch to subsystem 1 and from the other branch to subsystem 2. Since the sequence in one branch is purely sequential, the resulting subsystem language is supposed to be small which is positive for a quick convergence of the observed subsystem language. A possible strategy to determine subsystems from a Grafcet is thus to determine substructures with a sequential execution like in one of several parallel branches. Controller I/Os which are activated (outputs) or checked (inputs) in a sequential substructure can be grouped in a subsystem.

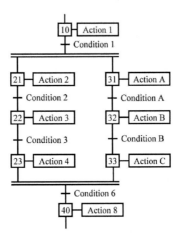

Figure 5.2: Example for a Grafcet

A second possibility to determine subsystems is to use human observations of the given system. For the case study system of figure 4.13 (page 85) for example, it can be observed that even during the parallel treatment of three work pieces the behavior of a machine tool and the conveyor in front of it is purely sequential. If the first machine tool and its conveyor are observed in detail, it can be seen that a new action in the subsystem only starts if the former action already finished: The drilling tool is only moved down if the work piece has stopped at the according position. This means that the conveyor must be stopped before the drilling tool is started. The conveyor is only reactivated when the drilling tool finished its task and has been moved back in its initial station. Hence, the according controller I/Os should be grouped in one subsystem.

Figure 4.13 shows that the position sensors between two conveyors are always associated with two subsystems. This is a consequence of the considerations from section 5.1. The value of the sensor between two conveyors can be influenced by the actuators of both conveyors: The conveyor in the left of the sensor can move a work piece to the sensor and the conveyor in the right of the sensor can transport a work piece away. Since it is advantageous for diagnosis purposes to cover causal relations in the subsystems, the sensors are associated to both subsystems.

Determining appropriate subsystems manually always requires expert knowledge. For large systems this knowledge is not easily available. The same problem like described in section 2.4.1 arises: System experts with the necessary knowledge to divide the system in subsystems are often not familiar with discrete event dynamics and concepts like 'concurrency'. Hence, it is difficult to use their knowledge for the partitioning approach. The second problem is that the necessary system analysis is a time consuming and expensive task.

To apply the distributed identification approach even if the system cannot be partitioned manually, in the next section an automatic approach working on the observed system language is introduced. It is capable of working with very few system knowledge but also allows integrating expert knowledge concerning causal relations to improve the diagnosis capability of the identified models.

5.3 Automatic partitioning based on observed data

5.3.1 Analysis of the solution space

The problem of dividing a closed-loop DES in subsystems can be formalized as the determination of the I/O-mapping function $y(sys_t)$ from definition 25. The set $y(sys_t)$ contains the I/Os which are considered in the partial I/O vector of the t-th subsystem. In order to evaluate possible strategies to perform the system partitioning automatically, it is helpful to determine the size of the solution space. The question of how many different possibilities exist to divide a set of m controller I/Os in n different subsystems is a combinatorial problem. If the closed-loop DES has three controller I/Os $\{IO_1, IO_2, IO_3\}$ which are to be partitioned in two non-empty, disjunct sets (subsystems), the following three possibilities exist:

$$y(sys_1) = \{IO_1, IO_2\} \qquad y(sys_1) = \{IO_1, IO_3\} \qquad y(sys_1) = \{IO_1\}$$
$$y(sys_2) = \{IO_3\} \qquad y(sys_2) = \{IO_2\} \qquad y(sys_2) = \{IO_2, IO_3\}$$

The number of possible disjunct partitions is given by the Stirling number of the second kind (Riordan, 2002):

$$S(m,n) = \frac{1}{n!} \sum_{i=0}^{n} (-1)^i \binom{n}{i} (n-i)^m \tag{5.1}$$

with m denoting the number of I/Os according to definition 24 and n denoting the number of non-empty subsystems. In case of 30 I/Os and three disjunct partitions (three subsystems and three identified partial automata), this leads to $2.06 \cdot 10^{14}$ possible solutions. Only a few of these possible solutions represent an appropriate partitioning for the distributed identification approach. The size of the solution space necessitates a search technique that avoids testing each possible solution. In the following sections a heuristic optimization approach is used to find an optimal solution. Like in the case of genetic algorithms the optimization technique does not test each possible solution but only some subspace of the complete solution space.

Using optimization techniques for identification of discrete event systems is a known approach as explained in section 2.4.2 and 2.4.3. The new idea in this work is to combine an optimization approach with a classical algorithmic identification method. The optimization approach is used for preprocessing of the observed language. The construction of the model itself is carried out with the algorithmic approach of chapter 4. Since the determination of the model structure is not performed with the optimization approach, this computationally heavy task is solved more efficiently than in existing optimization approaches for DES identification. It will be shown that the approach delivers results for real world systems in a reasonable time.

Using an optimization approach necessitates the formulation of optimization criteria. Since the aim is to minimize the internal concurrency in the subsystems, special measures are introduced in the next section. They formalize two phenomena which are strongly related to concurrency and which can be observed when analyzing the observed data. The aim is to use the measures as optimization criteria to choose appropriate subsystems.

5.3.2 Manifestation of concurrency in the observed system language

Existing approaches

Analyzing observed event sequences to discover concurrency is a research area in the field of software (re-)engineering. A possible application is the development of an information system for a complex business process. The classical way is to have the system analyzed by a designer building a model coding the causal relations and concurrent structures of the business process. Like in the case of closed-loop DES, modeling is an expensive and laborious task. Several attempts have been made to systematically analyze observed event sequences and to determine causal and concurrent structures.

In (Maruster et al., 2003) five metrics for the analysis of event sequences are introduced. The most interesting ones are the causality metric CM and the so called XY and YX metrics. The causality metric CM is based on the assumption that when first event x and shortly later event y occur it is possible that x causes the occurrence of y. CM is calculated by parsing the collected event sequences. If y occurs n events after x, CM is incremented with the factor δ^n, where δ is the so called causality factor ($\delta \in [0.0, 1.0]$). The factor δ is a free parameter that must be chosen upon a priori knowledge. If the event order is inverse (y occurs n events before x), CM is decreased by δ^n. After processing each event sequence, CM is divided by the minimum of the overall frequency of x and y (number of event occurrences). If CM is near to 0, there is no causal relation between x and y. Hence, they can be considered as concurrent.

The XY and YX metric is used to distinguish so called exclusive and parallel relations. A relation is called exclusive if the event x never appears directly before or directly after the event y. The XY and YX metrics are defined as

$$XY = \frac{|X > Y|}{\min(|X|, |Y|)} \tag{5.2}$$

$$YX = \frac{|Y > X|}{\min(|X|, |Y|)} \tag{5.3}$$

$|X > Y|$ ($|Y > X|$) denotes the number of occurrences of the string xy (yx) and $|X|$ ($|Y|$) denotes the number of occurrences of x (y). If XY and YX are near to zero the events x and y are supposed to be exclusive. If *both* metrics have relatively high values at the same time, x and y are supposed to be concurrent.

The metrics of (Maruster et al., 2003) allow comparing two arbitrary events and to get information about their causal or concurrent relation. A problem is that each event has to be compared with each other event. If many events and long sequences are to be treated, calculating the metrics can become computationally difficult.

A very similar approach is presented in (Cook et al., 2004). Among other metrics, the authors use the notion of entropy to measure the information carried by an event. The entropy is indicated by the so called conditional probability of occurrence:

$$P(S) = \frac{Occur(S)}{Occur(Prefix(S))} \tag{5.4}$$

where $Occur(S)$ denotes the number of occurrences of sequence S and $Prefix(S)$ denotes the sequence S with the last event removed. The entropy can be measured with $P(S)$ as follows: If the event y always occurs after the event x, $P(xy) = 1.0$ and for all other events ε, $P(x\varepsilon) = 0.0$. The direct following behavior of x is deterministic. Hence, the entropy is 0. If any other event occurs as often after x as it does y, the entropy increases until it reaches 1.0 indicating that the following behavior of x is completely random. In this case, $P(S)$ is very small (near 0). In (Cook et al., 2004), a special formula to precisely calculate the entropy based on $P(S)$ is given. It allows calculating the entropy based on observed event sequences. On the other hand, the entropy resulting from different concurrent model structures is calculated. An example is the so called fork structure. It describes the case of the Petri net in figure 5.3. After the occurrence of x, both y and z can occur due to the concurrent structure. Ideally, yz and zy occur with the same frequency ($P(xyz) = P(xzy)$) in the observed event sequences. For this case, the entropy is calculated with a special formula derived in (Cook et al., 2004). If the entropy calculated with $P(xyz)$ and $P(xzy)$ determined from the analysis of the observed event sequences is near to the pre-calculated value it can be concluded that a fork-style concurrency is present.

Although both approaches deliver some useful metrics to measure concurrency in event traces, they are not appropriate for closed-loop DES. The main problem is that they cannot directly be applied to I/O vector sequences. The result would be an information about the degree of concurrency of several I/O vectors. This is not the necessary information for partitioning of the closed-loop DES. The interesting question is to determine controller I/Os which behave concurrently. This would necessitate a new definition of the event sequences for closed-loop DES. A straight forward approach

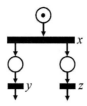

Figure 5.3: Example for a fork style concurrency

would be to define the change in value of a single I/O as an event and to generate event sequences from I/O vector sequences. Since more than one I/O can change its value between two I/O vectors, the result would be events occurring at the same time. Neither the approach of (Maruster et al., 2003) nor the approach of (Cook et al., 2004) is able to process this kind of event sequences. The second problem with the described approaches is that both deliver a variety of metrics (e.g. the comparison with different concurrent structures like the one shown in figure 5.3) which have to be analyzed in *their entirety*. This makes it difficult to formulate an optimization criterion since different measures have to be considered in parallel. Hence, it necessary to introduce new measures which are appropriate for closed-loop DES.

Language growth

The first measure is directly related to the precondition for the completeness of the identified distributed models. The distributed model is complete if for each of the subsystems sys_i, $L_{Ident,sys_i}^{k+n} \supseteq L_{Orig,sys_i}^{k+n}$ holds (theorem 6). Assumption 5 says that this can be stated if the according language converges to a stable level.

Definition 37 (Observed language up to the h-th system evolution). $L_{Obs,sys_i}^{n,h}$ denotes the observed language of length n from subsystem sys_i constructed according to definition 20 on the basis of the first h (of p) observed sequences: $\Sigma = \{\sigma_1, \ldots, \sigma_h\}$. The observed sequences are built on the basis of the partial I/O vectors of subsystem sys_i according to definition 26.

If the observed language does not converge to a stable level, the following equation holds:

$$|L_{Obs,sys_i}^{n,h}| - |L_{Obs,sys_i}^{n,h-1}| > 0 \qquad (5.5)$$

Since the h-th observed sequence contains at least one partial I/O vector that has not been seen before, the cardinality of the language $L_{Obs,sys_i}^{n,h}$ is larger than the cardinality of $L_{Obs,sys_i}^{n,h-1}$. This effect allows formulating a measure for the convergence of the observed language for an I/O partitioning given by the I/O mapping function y.

$$J_1(y) = \frac{1}{N_{sys}} \sum_{\forall sys_i} \sum_{h=2}^{p} (\sqrt{h}(|L_{Obs,sys_i}^{n,h}| - |L_{Obs,sys_i}^{n,h-1}|)) \qquad (5.6)$$

with p denoting the overall number of observed sequences an N_{sys} denoting the number of subsystems (according to definition 25 on page 66). If for a given partitioning y, there is a high degree of concurrency in one of the subsystems sys_i, then the according language growth will be important and thus lead to high values of $J_1(y)$. In equation 5.6 the running index h starts at $h = 2$. The result is that the first difference build in the measure is the language growth between the first and the second sequence. The first sequence can add an arbitrary number of new words to the language without having an influence on the measure. This reflects the fact that in the case of perfectly sequential subsystem behavior it is possible to observe the complete language during the first system evolution. This effect has already been shown for the subsystems chosen upon a priori knowledge in the case study system (see figure 4.14 on page 85). Starting the running index h at $h = 2$ avoids the measure being influenced by the new words in the first sequence which define the minimum of the observed language necessary to identify perfectly sequential subsystem behavior. The factor $\sqrt{h}$ adds more weight to new words occurring at later observed system evolutions. The earlier an observed language converges to a stable level, the more confidently can be stated that it has been completely observed. Hence, words leading to an early language growth are less negative then words leading to a growth during the last observed evolutions. During tests with the optimization method, the factor $\sqrt{h}$ delivered better results than simply multiplying with h. By division with the number of subsystems, the optimization criterion is normalized. For the approach presented in section 5.3.3 this facilitates determining different numbers of subsystems with the same set of optimization parameters.

Since the measure of equation 5.6 will be used as an optimization criterion, it should be possible to calculate it as fast as possible to increase the efficiency of the optimization approach. The measure basically consists of building and counting the observed words of length $1..n$. According to definition 20, a new word in the word set W_{Obs}^1 automatically leads to new words in the sets W_{Obs}^j $\forall 1 < j \leq n$. More generally, a new word in $W_{Obs}^{m<n}$ automatically leads to new words in the sets W_{Obs}^j, $\forall m < j \leq n$. Since it is not possible to have new words w^m with $m < n$ in L_{Obs}^n without leading to new words w^n, it is sufficient to consider the sets $W_{Obs,sys_i}^{n,h}$ and $W_{Obs,sys_i}^{n,h-1}$ in equation 5.6:

$$\widetilde{J}_1(y) = \frac{1}{N_{sys}} \sum_{\forall sys_i} \sum_{h=2}^{p} (\sqrt{h}(|W_{Obs,sys_i}^{n,h}| - |W_{Obs,sys_i}^{n,h-1}|)) \tag{5.7}$$

This avoids building and counting words with length lower than n. Since this saves computation time, $\widetilde{J}_1(y)$ is more appropriate for the optimization approach. If the observed language of the case study system is considered (without partitioning in subsystems), the calculation of $J_1(y)$ takes 4400ms whereas the calculation of $\widetilde{J}_1(y)$ takes only 2500ms for L_{Obs}^3 and W_{Obs}^3 respectively. Since the calculation has to be carried out up to several thousand times during optimization, this gain in calculation time is interesting.

Branching degree

The second measure is related to the structure of an automaton identified on the basis of observed system data. Figure 5.4 shows an example. In the right part of the figure a part of the reachability graph of a Petri net is shown. It can be seen that the three parallel branches of the Petri net lead to a state with three leaving transitions (state 2). Since after the occurrence of a the events b or c or d can be produced, the reachability graph contains three following transitions in state 2. If event b was produced, the automaton is in state 3. In this state both events c or d can be produced due to concurrency. Hence, state 3 has two leaving transitions. This shows that concurrency typically leads to several possible behaviors which are reflected by states with several leaving transitions in the reachability graph of an underlying Petri net.

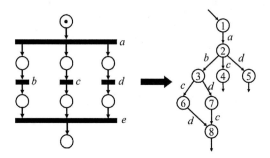

Figure 5.4: Part of a reachability graph

If for a given closed-loop DES an NDAAO is identified with algorithm 1, the resulting automaton can be seen as an approximation of the reachability graph of a Petri net representing the considered system. Hence, concurrency in the system is represented by states with several leaving transitions. As an example we consider the following three observed sequences:

$$\sigma_1 = (\begin{pmatrix} 1 \\ 0 \\ 0 \\ 1 \end{pmatrix}, \begin{pmatrix} 0 \\ 1 \\ 0 \\ 1 \end{pmatrix}, \begin{pmatrix} 0 \\ 1 \\ 1 \\ 0 \end{pmatrix}); \sigma_2 = (\begin{pmatrix} 1 \\ 0 \\ 0 \\ 1 \end{pmatrix}, \begin{pmatrix} 1 \\ 0 \\ 1 \\ 0 \end{pmatrix}, \begin{pmatrix} 0 \\ 1 \\ 1 \\ 0 \end{pmatrix}); \sigma_3 = (\begin{pmatrix} 1 \\ 0 \\ 0 \\ 1 \end{pmatrix}, \begin{pmatrix} 0 \\ 1 \\ 1 \\ 0 \end{pmatrix})$$

Applying algorithm 1 with $k = 1$ to the three sequences leads to the automaton in figure 5.5 on the left. It can be seen that the initial state has three leaving transitions that are necessary to reproduce the different observed sequences. In the example we assume that the global system consists of two concurrent subsystems: The first subsystem consists of the first two I/Os of the I/O vector and the second subsystem consists of the remaining two I/Os. It can be seen that the I/Os of the subsystems change their values concurrently leading two three possible following behaviors of the first state. In one following behavior the first subsystem leads to earlier I/O changes (σ_1). In the next case, the second subsystem evolves faster (σ_2) and in the third case, both systems evolve simultaneously (σ_3).

If for each subsystem an own partial automaton is identified, the resulting automata do not have states with several leaving transitions as depicted in figure 5.5 on the right. Since the subsystem behavior is sequential, the initial states of the partial automata have only one leaving transition. This shows that reducing concurrency has a direct effect

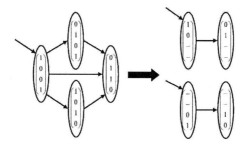

Figure 5.5: Result of monolithic and distributed identification

on the identified automata. To measure this effect, the following definition introduces the branching degree:

Definition 38. The branching degree BD of the NDAAO_{sys_i} identified for subsystem sys_i is defined as

$$BD(\text{NDAAO}_{sys_i}) = \sum_{\forall x \in X} \begin{cases} 0 & \text{if } |f(x)| \leq 1 \\ |f(x)| - 1 & \text{if } |f(x)| > 1 \end{cases}$$

The function (conditionally) counts for each state of the NDAAO identified for subsystem sys_i the leaving transitions. Only states with more than one leaving transition ($|f(x)| > 1$) contribute to this measure since this represents possible concurrent behavior. For states with several leaving transitions, we subtract one from the number of transitions. This is done since one following transition in a state does not represent concurrency. In the example of figure 5.5, the monolithic model has a BD of 2 since only the first state contributes to the measure and has three leaving transitions. Each partial automaton on the right of figure 5.5 has a BD of 0 since due to the purely sequential subsystem behavior each state has at most one leaving transition.

The branching degree is an absolute measure since it is not normalized to the size of the automaton. It only considers states with several possible following behaviors but ignores states with only one leaving transition. A normalization to the size of the automaton (e.g. to the state space) could lead to a situation where the branching degrees of two automata with a differing number of states are not the same even if both automata contain the same number of states with more than one leaving transition. In this case, the automata reflect the same degree of concurrency and should thus lead to the same value for BD. Hence, the automata in figure 5.6 have the same branching degree ($BD = 2$) although one automaton has more states. The long state trajectory before state 7 is ignored by the measure since it reflects sequential behavior. The same holds for the states A and B in the second automaton. The non-normalization is an important property of the measure since the optimal solution for I/O partitioning can lead to subsystems resulting in automata of different size.

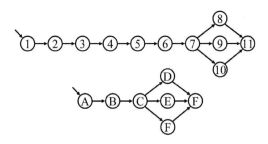

Figure 5.6: Example for the non-normalization of the branching degree

A measure for the concurrency of a given I/O partitioning y can be calculated by summing up the branching degrees of partial NDAAO identified for the subsystems:

$$J_2(y) = \sum_{\forall sys_i} (BD(\text{NDAAO}_{sys_i}) \qquad (5.8)$$

with NDAAO_{sys_i} denoting the automaton identified with a given parameter k for subsystem sys_i.

It is possible to have $BD > 0$ and thus $J_2(y) > 0$ although there is no concurrency in the system. If several 'decisions' exist in the system (e.g. large OR small work piece), it is possible to have several following states in a given identified automaton to reflect the different possible conditional (and not concurrent) behaviors. The same effect happens if the identification parameter k is not chosen large enough as explained in section 3.4.3. This can lead to states in the identified automaton approximating several closed-loop DES states. Such an automaton state can have several leaving transitions to represent each possible following behavior of the represented DES states. During the optimization approach presented in the next sections, the aim is to find a y leading to a *minimized* $J_2(y)$. It is not necessary to reach branching degree of zero. Hence we can cope with $BD_{min} > 0$.

5.3.3 A heuristic optimization approach: Simulated Annealing

As explained in section 5.3.1, partitioning a given closed-loop DES into appropriate subsystems is a combinatorial problem. With the measures introduced in section 5.3.2 it is possible to evaluate if a possible partitioning y represents a useful solution. The immense size of the solution space (equation 5.1) makes it impossible to systematically evaluate each solution. A possible way to solve combinatorial problems without analyzing the complete solution space is to use so called metaheuristic approaches like simulated annealing (Michalewisz and Fogel, 2000). Simulated annealing is an optimization algorithm inspired by the annealing process in metallurgy. In metallurgy, a possible way to influence material properties is to heat them to above their crystallization temperature. Due to the heat, atoms become unstuck from their initial positions (corresponding to a local minimum of the internal energy) and wander randomly to other positions with other degrees of energy. Slow cooling gives them the chance to reach configurations with lower internal energy than the initial one. If the atoms reach positions with a lower internal energy, this relieves internal stresses which improves the cold working properties of the metal. The simulated annealing algorithm simulates this process. If it is well parameterized, it is more efficient than classical gradient descent methods due to its capability of leaving a local minimum.

Algorithm 6 shows a possible implementation for simulated annealing. It is a slightly modified version of the algorithm in (Michalewisz and Fogel, 2000). In the first step, the metaphorical temperature is set to the initial temperature T_0. The initial temperature T_0 is a free variable which must be chosen by the user. In the next step, a first solution for the I/O mapping function y_c (c for current) is chosen. It randomly assigns all I/Os to subsystems. A description of this function will be given in the next two sections. In line 3, the main optimization loop starts. First, a new solution y_{new} is chosen. A detailed implementation for choosing y_{new} on the basis of y_c is given in the next section. In the next step, the new solution is compared to the current solution. The function $J(y)$ implements one of the fitness functions $\tilde{J}_1$ or J_2 from the former section. If the new solution y_{new} leads to a lower value of the optimization criterion, y_{new} becomes the new current solution y_c. In this case, the optimization algorithm works like a gradient descent method. If the new solution is not better than y_c it is decided *probabilistically* if the new solution nevertheless becomes the current one. This helps to avoid getting stuck in a local minimum. The probabilistic choice is based on a calculation considering the current temperature and the difference between the fitness of y_c and y_{new}. In line 7, $J(y_c) - J(y_{new})$ can only be negative or zero, since $J(y_c) \leq J(y_{new})$. The difference is divided by the current temperature. Although the temperature is decreased during each repetition of the optimization loop, it keeps a positive value. Hence, the difference divided by T is always a negative number or zero leading to the exponential function always returning a value in the interval $(0, 1]$. A given difference $J(y_c) - J(y_{new})$ leads to a larger result of the exponential function in an early cycle of the optimization algorithm (large T) than in a later cycle with a decreased T. The new solution becomes the current one if the result of the exponential function is larger than a randomly chosen number of the interval $[0, 1)$. This makes accepting a worse solution y_{new} as the new solution y_c more probable if the difference $J(y_c) - J(y_{new})$ is small and the temperature is still high. In line 10, the temperature is decreased with the cooling rate CR (with $0 < CR < 1$). The algorithm stops if the predefined lower bound of the temperature T_{min} is reached.

Since it is possible to accept a worse solution y_{new} as the current solution, it is reasonable to save each analyzed solution with the according fitness value in a list. It is possible that the probabilistic nature of the algorithm leads to a final solution which is not the best one analyzed so far. In this case, the best solution from the list can be taken.

An important part of the optimization algorithm is the determination of possible solutions. In the next two sections, two approaches for determining a new I/O partitioning based on a current solution are presented.

Algorithm 6 Simulated annealing

Require: initial temperature T_0, minimum temperature T_{min}, cooling ratio CR

1: $T := T_0$ initialize temperature
2: select current solution y_c at random
3: **repeat**
4:     select a new solution y_{new}
5:     **if** $J(y_c) > J(y_{new})$ **then**
6:         $y_c := y_{new}$
7:     **else if** $random([0,1)) < e^{\frac{J(y_c)-J(y_{new})}{T}}$ **then**
8:         $y_c := y_{new}$
9:     **end if**
10:    $T := T * CR$
11: **until** $T < T_{min}$

5.3.4 Minimal knowledge solution

The question of how to select possible solutions in algorithm 6 is closely related to the degree of available a priori knowledge. First, we assume that the available knowledge only allows defining the necessary number of subsystems. If this knowledge is not available, the number of subsystems can be increased bit by bit until the optimization algorithm admits an acceptable solution.

At the beginning of algorithm 6, an initial I/O partition has to be determined as first solution y_c. A common way to get a first solution for simulated annealing is to determine it with a random approach if no better knowledge is available. Algorithm 7 shows the implementation for the partitioning problem. First, it assigns empty sets to each subsystem. Via the *random*-function, a subsystem is selected randomly from the set of all subsystems. This allows assigning each controller I/O IO_i randomly to one of the subsystems. The algorithm terminates if each subsystem contains at least the predefined minimal number of I/Os.

Algorithm 7 Determine initial solution with minimal knowledge

Require: number of subsystems N_{sys}, global I/O vector u, minimal number of I/Os in a subsystem $minIO$

1: **repeat**
2:    $\forall t \in 1, \ldots, N_{sys} : y(sys_t) := \{\}$
3:    **for all** $IO_i \in u$ **do**
4:       $randSys := random(\{sys_1, \ldots, sys_{N_{sys}}\})$
5:       $y(randSys) := y(randSys) \cup IO_i$
6:    **end for**
7: **until** $\forall t \in 1, \ldots, N_{sys} : |y(sys_t)| \geq minIO$

Apart from determining an initial solution it is also necessary to continuously construct new solutions when applying the optimization algorithm. Although simulated annealing has some non-deterministic characteristics, it is not reasonable to take completely random samples from the solution space. In the case of finding appropriate subsystems it can be assumed that having a good solution, it is more efficient to look for a better solution in the close 'neighborhood' of the original solution than considering the complete search space. If for example 90% of the I/Os have been assigned appropriately in a given solution y_c, it is more likely to get a good solution y_{new} by only altering the assignement of a few I/Os than rejecting the complete solution and determining a new one e.g. with algorithm 7. Algorithm 8 shows a way to alter an existing solution. First, the old solution is copied to the new one. Then, a predifined number of I/Os is randomly chosen. The solution difference sD is a paramter which defines how close the new and the old solution are. It defines the number of I/Os for which the according subsystem will be altered. Each randomly selected I/O is erased from its subsystem. In lines 6 and 7, a new subsystem $randSys$ is chosen randomly and the selected I/O is added to $randSys$. If the procedure leads to a forbidden solution (with at least one subsystem consisting of less than $minIO$ I/Os) the algorithm starts again.

Algorithm 8 Determine next solution with minimal knowledge

Require: Existing solution y_c, number of subsystems N_{sys}, global I/O vector u, minimal number of I/Os in a subsystem $minIO$, solution difference sD

1: **repeat**
2:    $y_{new} := y_c$
3:    $IOSet := \{sD$ randomly chosen I/Os from the global I/O vector$\}$
4:    **for all** $IO_i \in IOSet$ **do**
5:       $y_{new}(sys)|IO_i \in y_{new}(sys) : y_{new}(sys) := y_{new}(sys) \backslash IO_i$
6:       $randSys := random(\{sys_1, \ldots, sys_{N_{sys}}\})$
7:       $y_{new}(randSys) := y_{new}(randSys) \cup IO_i$
8:    **end for**
9: **until** $\forall t \in 1, \ldots, N_{sys} : |y_{new}(sys_t)| \geq minIO$

It is obvious that the choice of the solution difference parameter sD has a direct influence on the performance of the optimization approach. If it has a too small value, algorithm 8 is not able to determine a new solution varying enough to leave a local minimum. If sD is too large, it is difficult to come close to a minimum since the new solutions will be situated far away in the search space. Hence, it is crucial to properly choose sD. It can be necessary to perform different optimization runs with various values for sD to get a sufficiently good solution. The practical experience gained throughout this work showed that $sD = 3$ is a good starting value for systems with up to 200 controller I/Os.

5.3.5 Integrating knowledge about causal relations

The minimal knowledge approach presented in the former section does not allow considering any structural information of the closed-loop DES. In many practical cases, some limited system knowledge is available. The knowledge considered in this section is the (possibly partial) knowledge of causal relations of actuators and sensors in the plant. As described in section 5.1, covering causal relations makes a model more appropriate for fault diagnosis purposes. It is thus preferable to integrate this knowledge if available. The idea of the algorithms in this section is to assign controller I/Os always together with causally influenced I/Os to subsystems.

The knowledge about causal actuator sensor relations belongs to the system expert domain. Since a system expert is usually not familiar with DES modeling or identification techniques, it is necessary to develop a tool to formalize this knowledge. An intuitive way to capture the knowledge is to enumerate for each actuator the sensors which can be influenced by it. In terms of the controller I/O vector, this can be implemented by a function assigning a set of controller inputs to each controller output:

Definition 39 (Causal I/O map). The function $CausalMap : IO \rightarrow 2^{IO}$ assigns to each controller input a set of controller outputs.

If no other information is given, we assume that for the initial implementation of the function it holds $\forall IO_i \in u : CausalMap(IO_i) = \{\}$ with u denoting the controller I/O vector. To capture the causal actuator sensor relations, a system expert has to fill the function $CausalMap$ like depicted in figure 5.7. In the figure on the left, only the output part of the controller I/O vector is shown whereas on the right only the inputs are given. The expert has to assign the relation of outputs and inputs. Setting a given controller output (and thus starting or stopping an actuator) at different internal states of a closed-loop DES may not always influence the same sensors (controller inputs). It is nevertheless preferable to enumerate *all* inputs which are potentially influenced by setting a controller output in order to capture the actuator sensor relations as completely as possible. An example for such an actuator sensor relation is a conveyor with three position sensors like in figure 3.6 on page 52. If the conveyor is transporting a work piece, it depends on the actual work piece position which position sensor is influenced next. Since the actuator of the conveyor potentially influences each position sensor, the according controller output should be related to each input leading to a position sensor. Like shown in figure 5.7, it is thus possible to assign several inputs to a single output. It can also be seen that some controller inputs are influenced by several controller outputs. On the other hand it is possible that the expert cannot determine a set of influencing outputs for each input or vice versa.

Another aspect of building the causal I/O map is covered by the following assumption:

Assumption 6. It is assumed that $\forall IO_i \in u$ the following condition holds:

$$\text{if } CausalMap(IO_i) \neq \{\} \text{ then } \forall IO_j \in u : IO_i \notin CausalMap(IO_j)$$

By assumption 6 it is made clear that each I/O influencing at least one other I/O is not itself influenced. The assumption is always met if the only possible influence captured in $CausalMap$ is from controller outputs to controller inputs like in figure 5.7.

Considering the information captured in the causal I/O map, it is necessary to modify the algorithms determining initial and next solutions. First, a modified version of

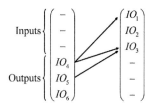

Figure 5.7: Example for the assignments in a causal map

algorithm 7 to determine an initial solution is presented with algorithm 9. At the beginning of the algorithm, empty I/O sets are assigned to each subsystem. In line 3 the first inner loop starts. It considers each controller I/O which influences at least one other I/O. In line 4 and 5 the according I/O and the I/Os which are influenced by it are assigned to a randomly chosen subsystem. If the causal relations are not completely known, some controller outputs exist with $CausalMap(IO_i) = \{\}$. It is also possible that some controller inputs exists which are not part of any actuator sensor relation covered in the causal I/O map. The expression in line 7 determines this group of controller I/Os. In line 8 and 9, the considered I/Os are assigned to a randomly chosen subsystem. The algorithm is repeated until each subsystem contains at least the predefined minimum number of I/Os.

Algorithm 9 Determine initial solution with a causal I/O map

Require: number of subsystems N_{sys}, global I/O vector u, minimal number of I/Os in a subsystem $minIO$, causal I/O map $CausalMap$

1: **repeat**
2:    $\forall t \in 1, \ldots, N_{sys} : y(sys_t) := \{\}$
3:    **for all** $\{IO_i \in u | CausalMap(IO_i) \neq \{\}\}$ **do**
4:       $randSys := random(\{sys_1, \ldots, sys_{N_{sys}}\})$
5:       $y(randSys) := y(randSys) \cup IO_i \cup CausalMap(IO_i)$
6:    **end for**
7:    **for all** $\{IO_i \in u | CausalMap(IO_i) = \{\} \wedge IO_i \notin (CausalMap(IO_j)\forall IO_j \in u)\}$ **do**
8:       $randSys := random(\{sys_1, \ldots, sys_{N_{sys}}\})$
9:       $y(randSys) := y(randSys) \cup IO_i$
10:   **end for**
11: **until** $\forall t \in 1, \ldots, N_{sys} : |y(sys_t)| \geq minIO$

The next step is to modify algorithm 8 to determine a next solution during optimization. Algorithm 10 follows the same principles like algorithm 8 but also considers the causal I/O map. At the beginning of the algorithm, the current solution is copied to the new one. Then, an I/O set is built containing each I/O which is *not* causally influenced by any other I/O. If assumption 6 is met, the set is built such that each I/O is either directly part of it or it is causally influenced by some I/O in $IOSet$. From this set sD I/Os are randomly chosen. sD is the solution difference introduced in the former section. From line 5 to line 9, the randomly selected I/Os and the according causally influenced I/Os (defined by the causal I/O map) are taken from their initial subsystem. Together with their influenced I/Os, they are added to a new randomly chosen subsystem. The algorithm stops if each subsystem has at least the predefined minimum number of I/Os.

Algorithm 10 Determine next solution with a causal I/O map

Require: Existing solution y_c, number of subsystems N_{sys}, global I/O vector u, minimal number of I/Os in a subsystem $minIO$, solution difference sD, causal I/O map $CausalMap$

1: **repeat**
2:    $y_{new} := y_c$
3:    $IOSet := \{IO_i \in u | \forall IO_j \in u : IO_i \notin CausalMap(IO_j)\}$
4:    $IOSubSet := \{sD \text{ randomly chosen I/Os from } IOSet\}$
5:    **for all** $IO_i \in IOSubSet$ **do**
6:       $y_{new}(sys) | IO_i \in y_{new}(sys) : y_{new}(sys) := y_{new}(sys) \backslash (IO_i \cup CausalMap(IO_i))$
7:       $randSys := random(\{sys_1, \ldots, sys_{N_{sys}}\})$
8:       $y_{new}(randSys) := y_{new}(randSys) \cup IO_i \cup CausalMap(IO_i)$
9:    **end for**
10: **until** $\forall t \in 1, \ldots, N_{sys} : |y_{new}(sys_t)| \geq minIO$

In difference to algorithm 8, algorithm 10 does not only transfer single I/Os to other subsystems. If an I/O IO_i is given to a new subsystem, any I/O influenced by IO_i (which consequently is part of $CausalMap(IO_i)$) is also transfered. The algorithm assures that causal relations are maintained when exploring the solution space. As explained in section 5.1, I/Os with a causal relation are typically not concurrent. Hence, the presented approach is usually not an obstacle of finding subsystems with low internal concurrency but a help.

Apart from the advantages of capturing causal relations in the subsystems explained is section 5.1, the approach presented in this section has another positive impact on diagnosis: If several I/Os are influenced by more than one other I/O, the resulting subsystems can have overlapping I/Os. Since overlapping I/Os automatically restrict the combined automata network behavior (and thus the exceeding network behavior), they have a positive influence on the fault detection capability of an identified distributed model.

5.4 Partitioning of the case study system

5.4.1 Evaluation criteria

Before the optimization approach to divide a closed-loop DES can be evaluated with the case study system, it is necessary to define evaluation criteria. The aim is to provide means to systematically compare an automatically determined set of subsystems with a predefined 'good' solution. This makes use of the fact that for the case study system the necessary expert knowledge to find an appropriate solution is available. The better the similarity of the optimization results of the case study system to the predefined solution is, the more can be trusted in the method if it is applied to systems with few available knowledge. The knowledge of a good partitioning for the case study system is thus only used for evaluation purposes and not during the automatic determination of the partitioning.

The similarity of an automatically found solution to a predefined one can be given by two criteria:

1. Absolute number of controller I/Os shared by an automatically generated and a predefined subsystem.

2. Number of I/Os shared by an automatically generated and a predefined subsystem in relation to the number of I/Os in the predefined subsystem.

In the first criterion, the cardinality of the intersection of a predefined and an automatically generated subsystem is calculated:

$$EC_1(sys_{aut}, sys_{predef}) = |sys_{aut} \cap sys_{predef}| \tag{5.9}$$

with sys_{aut} and sys_{predef} denoting an automatically chosen and a predefined subsystem respectively. The higher EC_1 is, the more similar both subsystems are. This criterion only counts positive matches. Wrongly assigned I/Os are ignored but decrease the quality of the automatically found solution. The second criterion helps to overcome this problem since it relates the number of correctly assigned I/Os with the total number of I/Os in the predefined subsystem:

$$EC_2(sys_{aut}, sys_{predef}) = \frac{EC_1(sys_{aut}, sys_{predef})}{\max(|sys_{predef}|, |sys_{aut}|)} \times 100\% \tag{5.10}$$

Equation 5.10 delivers the proportion of correctly assigned I/Os. The following example shows how the criteria are used to compare two solutions y_{aut} and y_{predef} consisting of several subsystems. Table 5.1 shows two solutions y_{aut} and y_{predef} as well as the resulting values for the evaluation criteria. Comparing the predefined subsystem sys_{III} and the automatically determined sys_B for example shows that they share two I/Os ($EC_1 = 2$) leading to $EC_2 = 67\%$.

So far, the criteria have only been used to compare two given subsystems. The original aim of the evaluation criteria is to compare two solutions for the partitioning problem, each consisting of several subsystems. The comparison of two solutions necessitates combining the results of several subsystem comparisons. In table 5.1, the different possibilities to compare the subsystems of the predefined and the automatically determined solution are shown. To compare the two solutions it is for example possible to compare sys_I with sys_A, sys_{II} with sys_B and sys_{III} with sys_C. Overall, six possible ways to compare the subsystems of the both solutions exist. To determine the similarity of several subsystem pairs, the resulting values for EC_1 and EC_2 have to be combined. The combination of the first criterion (referred to as $\overline{EC_1}$) is the sum of the single values of the subsystem comparisons. For the first row in table 5.2 we get:

$$\overline{EC_1} = EC_1(sys_I, sys_A) + EC_1(sys_{II}, sys_B) + EC_1(sys_{III}, sys_C) = 0 + 1 + 0 = 1 \tag{5.11}$$

		Automatically generated				
		$y_{aut}(sys_A) =$ $\{IO_6, IO_9\}$		$y_{aut}(sys_B) =$ $\{IO_1, IO_3, IO_7, IO_8\}$		$y_{aut}(sys_C) =$ $\{IO_2, IO_4, IO_5\}$
		EC_1	EC_2	EC_1	EC_2	EC_1	EC_2
Predefined	$y_{predef}(sys_I) =$ $\{IO_1, IO_2\}$	0	0%	1	50%	1	50%
	$y_{predef}(sys_{II}) =$ $\{IO_3, IO_4, IO_5, IO_6\}$	1	25%	1	25%	2	50%
	$y_{predef}(sys_{III}) =$ $\{IO_7, IO_8, IO_9\}$	1	33%	2	67%	0	0%

Table 5.1: Example for the evaluation criteria

The combination of the second criterion (referred to as $\overline{EC}_2$) is calculated by the sum of EC_2 for the single subsystems. To keep a meaningful percentage (below 100%), the sum is divided by the number of subsystem comparisons. For the first line in table 5.2 we get:

$$\overline{EC}_2 = \frac{EC_2(sys_I, sys_A) + EC_2(sys_{II}, sys_B) + EC_2(sys_{III}, sys_C)}{3} = \frac{25\%}{3} = 8.3\%$$
(5.12)

Table 5.2 shows for all possibilities the resulting combined criteria. It can be seen that three comparison orders lead to the same $\overline{EC}_1$ value. To decide which one represents the best similarity, $\overline{EC}_2$ is considered. It can be seen that the last combination leads to the highest values in both $\overline{EC}_1$ and $\overline{EC}_2$. It is obvious that the comparison order has a strong influence on the result of comparing two solutions. In the following, we will always take the comparison order leading to the highest value in $\overline{EC}_1$. $\overline{EC}_2$ will also be given to serve as an additional indicator for the similarity (see e.g. figure 5.10). If several orders leading to the same $\overline{EC}_1$ exist, the one with the highest $\overline{EC}_2$ will be taken as basis for the comparison. This makes sure that the most similar predefined and automatically generated subsystems are compared.

Comparisons / Criteria	$\overline{EC}_1$	$\overline{EC}_2$
$sys_I \longleftrightarrow sys_A, sys_{II} \longleftrightarrow sys_B, sys_{III} \longleftrightarrow sys_C$	1	8.3%
$sys_I \longleftrightarrow sys_A, sys_{II} \longleftrightarrow sys_C, sys_{III} \longleftrightarrow sys_B$	4	39%
$sys_I \longleftrightarrow sys_B, sys_{II} \longleftrightarrow sys_A, sys_{III} \longleftrightarrow sys_C$	2	25%
$sys_I \longleftrightarrow sys_B, sys_{II} \longleftrightarrow sys_C, sys_{III} \longleftrightarrow sys_A$	4	44.3%
$sys_I \longleftrightarrow sys_C, sys_{II} \longleftrightarrow sys_B, sys_{III} \longleftrightarrow sys_A$	3	36%
$sys_I \longleftrightarrow sys_C, sys_{II} \longleftrightarrow sys_A, sys_{III} \longleftrightarrow sys_B$	4	47.3%

Table 5.2: Example for the combined evaluation criteria

In section 4.4, the system partitioning determined with a priori knowledge has been explained. It is depicted in figure 4.13 on page 85. In the following sections, this solution will be used as a reference for the automatically determined ones.

5.4.2 Partitioning with minimal knowledge

In this section the minimal knowledge approach from section 5.3.4 is applied to the case study system. For the determination of the optimization results in this chapter, the initial temperature T_0 in algorithm 6 was set to $T_0 = 1000$. The cooling ratio was set to $CR = 0.99$ and T_{min} was set to $T_{min} = 0.43$ leading to about 1000 iterations in the algorithm. The solution difference sD for algorithm 8 was set to $sD = 3$. The minimal number of I/Os in a subsystem $IOMin$ was set to 3.

In order to show that it is possible to get meaningful results even if the a priori knowledge does not allow predefining the necessary number of subsystems, the optimization approach has been carried out with two, three and four subsystems. Figure 5.8 shows the evolution of the optimization criterion $\tilde{J}_1(y)$ for different numbers of subsystems. The observed word sets of length $n = 2$ have been analyzed. The figures show $\tilde{J}_1(y)$ for the currently accepted solution. It can be seen that due to the stochastic nature of simulated annealing, the algorithm temporarily accepts worse solutions especially during early optimization cycles where the temperature is still high. At the end, each optimization converges to a stable value. It can be seen that $\tilde{J}_1(y)$ is decreased significantly from 123.98 to 19.44 when the number of subsystems is increased from two to three. Since further increasing the number of subsystems to four does not lead to a significant improvement of $\tilde{J}_1(y)$, it can be concluded that three is an appropriate number for the subsystems.

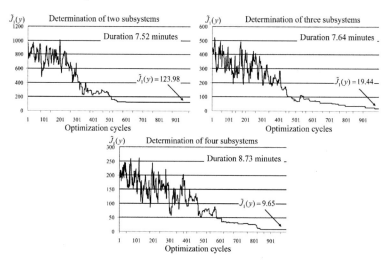

Figure 5.8: Determination of two, three and four subsystems with 'language growth'

Using the second optimization criterion leads to similar results. The necessary identification of partial NDAAOs was parameterized with $k = 2$. Figure 5.9 shows the evolution of the optimization criterion $J_2(y)$ for different numbers of subsystems. Again, it can be seen that the solution with three subsystems is better than the solution with two subsystems. Increasing the number of subsystems to four does not have a significant effect. The solution with three subsystems converges to $J_2(y) = 3$. This indicates that there are at maximum three states in the automata identified for the subsystems with more than one leaving transition. The partial automata thus represent the sequential subsystem behavior. In the next chapter it will become clear that a low number of leaving transitions significantly improves the fault isolation techniques based on the identified models.

The calculation of both optimization criteria can be parallelized. Once a new solution y has been determined during the optimization algorithm, each subsystem is analyzed separately. In case of $\widetilde{J}_1(y)$, the subsystem language is counted and in case of $J_2(y)$ the partial NDAAO are identified and analyzed. This makes it possible to calculate the results for the subsystems *in parallel*: For $J_2(y)$ it is e.g. possible to identify partial NDAAO for each subsystem independently and to sum up the branching degree results for the single subsystems. This allows exploiting the parallel computing capabilities of modern computers. Using an Intel®Core™2 Duo CPU (two CPU cores with 1.8GHz) leads to a reduction of the optimization time of about 30%. The duration of each optimization run can be seen in the figures.

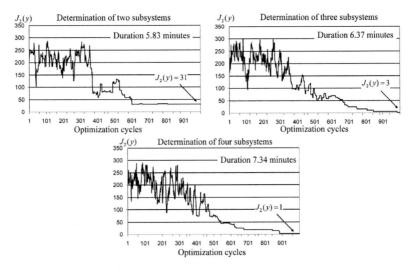

Figure 5.9: Determination of two, three and four subsystems with 'branching degree'

To analyze the similarity of automatically determined and predefined good solutions, figure 5.10 depicts the values $\widetilde{J}_1(y)$ and $J_2(y)$ in some selected optimization cycles. The results are taken from the optimization yielding three subsystems. The figure also shows the evaluation criteria defining the similarity to the predefined solution. The reference solution is the one shown in figure 4.13 on page 85. It can be seen that in both cases solutions occurring at later cycles are more similar to the reference solution than early ones (high values of $\overline{EC}_1$ and $\overline{EC}_2$). The figure shows that there is a strong relation between a low value of the optimization criteria and the similarity to the reference solution. This allows being confident about the potential of the optimization approach: Since the automatic approach delivers very useful results for the case study, it is very probable that it also performs in this way when similar systems are considered where no sufficient knowledge to a priori define a partitioning solution is available.

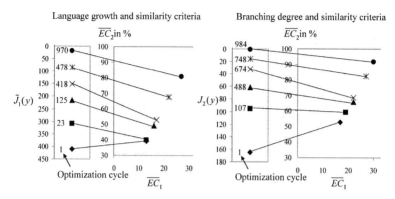

Figure 5.10: Evaluation of the results of the minimal knowledge approach

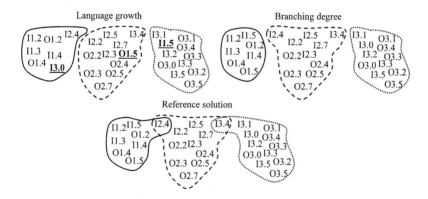

Figure 5.11: Result of the partitioning with the minimal knowledge approach

In figure 5.11, the resulting subsystem partition for both optimization criteria is shown. It can be seen that both criteria lead to very similar results which are close to the reference solution. The language growth criterion leads to some wrongly assigned I/Os (e.g. I3.0 in a subsystem with mainly I/Os from the first machine tool). Although using the branching degree criterion led to a better solution, this was not always the case. Starting the optimization algorithm several times always led to similar but mostly not equal solutions. The main difference between the automatically determined and the predefined solutions are the missing synchronization I/Os due to the next solution algorithm from section 5.3.4. Modifying algorithm 8 such that it assigns several I/Os to subsystems is possible from a theoretical point of view. Practical experiences with such a modified algorithm showed that this makes the solution space too large. It was not possible finding acceptable solutions. In the following section, it will be shown that using knowledge about causal relations helps to overcome this problem.

5.4.3 Partitioning with knowledge about causal relations

As shown in section 5.3.5, it is possible to use knowledge of causal actuator sensor relations to improve the results of the automated partitioning approach. First, it is necessary to define the causal map function according to definition 39 on page 106. Figure 5.12 shows the causal actuator sensor relations of the case study systems. It contains all I/Os from table 3.2 and figure 3.10 (page 59). It can be seen that some actuators like O1.4 (drilling motor) do not have an influence on any sensor. Others like O1.5 (conveyor in front of the drilling machine) have an influence on several controller inputs. Some sensors like I2.4 (position sensor between drilling and vertical milling machine) are influenced by several controller outputs.

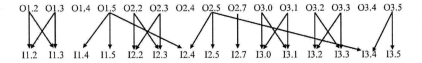

Figure 5.12: Causal actuator sensor relations of the case study system

Based on the actuator sensor map given in figure 5.12, the optimization approach is applied. Like in the former section, the initial temperature T_0 in algorithm 6 was set to $T_0 = 1000$. The cooling ratio was set to $CR = 0.99$ and T_{min} was set to $T_{min} = 0.43$ leading to about 1000 iterations in the algorithm. In this section, algorithm 10 is used to determine the next solution during the optimization. The solution difference sD was set to $sD = 3$ and $IOMin$ was set to 3.

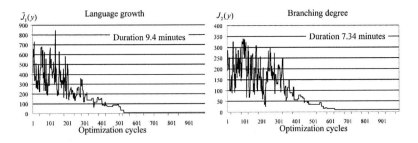

Figure 5.13: Evolution of the optimization criteria using causal knowledge (for the determination of three subsystems)

The qualitative results of determining two, three and four subsystems are similar to the ones shown for the minimal knowledge solution. Hence, only the case of finding three subsystems is considered. Figure 5.13 shows the evolution of the optimization criteria. Like in the former section, the optimization criteria are successfully minimized. It can be seen that using the branching degree criterion is slightly faster than using language growth.

Figure 5.14 shows that the approach using knowledge about causal relations also delivers results which are similar to the reference solution. The main difference to the minimal knowledge approach is that the resulting subsystems share several I/Os. Figure 5.15 shows the resulting partitions of the optimization. It can be seen that both criteria deliver close approximations to the reference solution. The only I/O which is wrongly assigned is O3.4. The solutions using causal knowledge both lead to the same synchronization I/Os.

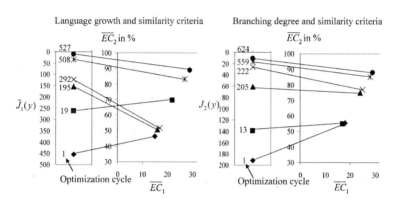

Figure 5.14: Evaluation of the results of the causal knowledge approach

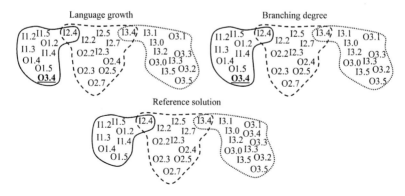

Figure 5.15: Result of the partitioning with causal knowledge

For the case study system the causal knowledge is completely available. The results from the former section show that even if this knowledge is not completely available, the optimization approach delivers useful results. Treating the case study system with the optimization approach showed that it is possible to obtain subsystems such that the precondition of theorem 6 is met. This makes sure that using the distributed models for fault diagnosis it is possible to significantly reduce the number of false alerts compared to the monolithic approach. The fact that the optimization approach delivered for the case study solutions which are very close to a solution defined based on a priori knowledge shows that the method has a considerable potential.

6 Fault Detection and Isolation

6.1 Fault detection and isolation with the identified fault-free monolithic model

6.1.1 Overview of the proposed method

After the chapters dealing with identification of fault-free system models in the first part of the work, chapter 6 addresses detecting and isolating faults based on the identified models. First, a method to work with the monolithic system model is presented in section 6.1. This approach is then adapted to the identified distributed models in section 6.2.

Figure 6.1 shows the online monitoring scheme using the monolithic system model. The model is run in an evaluator which tries to find a trajectory corresponding to the observed system behavior. If such a trajectory exists, there is no deviation between observed and modeled behavior. Hence, no fault is detected. If the evaluator is not able to reproduce the observed system behavior, a deviation exists and leads to fault detection. A deeper analysis of the deviation allows isolating the fault.

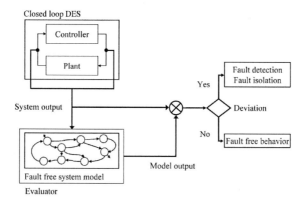

Figure 6.1: Online monitoring of closed-loop DES with an NDAAO

As explained in section 2.3, most of the known diagnosis methods for DES work with a model of the fault-free and the faulty behavior. Since in the case of the presented identified models only the the fault-free behavior is captured, a new approach has to be developed.

In continuous system theory, working with so-called residuals is a well-known technique to use fault-free system models for online fault diagnosis. (Isermann and Balle, 1997) define residuals as follows:

> A residual is a fault indicator, based on a deviation between measurements and model-based computation.

In the upper part of figure 6.2 this principle is depicted for continuous systems. The deviation between a measured signal evolution (dashed line) and the signal evolution computed using a model (solid line) is quantified. Analyzing this quantification, it is possible to detect and isolate faults in the observed system. Figure 6.2 also shows measurements of a DES and a (standard) automaton modeling the considered system. The observation of the DES is given by an event sequence as shown in the lower part of the figure. The expected event trace can be derived from the automaton modeling the system. The observed sequence a, b can be reproduced by the state trajectory 0, 4, 5. When event c is observed, the sequence is no longer reproducible with the given automaton. It can be seen that the observed event c was unexpected and the event d was missed. This makes it highly probable that one of these events is related to the detected fault. The presented approach focuses on the analysis of the difference between *measured* and *expected* DES behavior. Two generic fault symptoms will be considered in the following: Faults leading to *observed* but *unexpected* events and faults that lead to *missed* events. The idea is related to the principle of Parsimony of (Reiter, 1987) (see page 22) where diagnosis is understood as conjecture of a minimal set of components that has to be assumed to be faulty to explain the current observation.

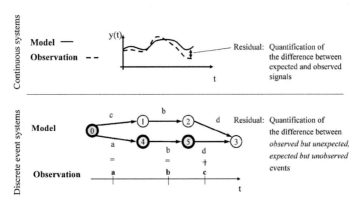

Figure 6.2: Residuals in continuous and in discrete event systems

The aim of the fault isolation method is to determine controller I/Os which are related to a faulty component. If for example a position sensor does not change its value due to a fault, we want to isolate the controller I/O related to this sensor. With this information, the maintenance personnel does not have to check each sensor and actuator but only those which are related to reported potentially faulty I/Os. Especially if large industrial facilities are considered, this significantly reduces the time to start the necessary repair actions.

The models defined so far create a language based on controller I/O vectors. To determine single I/Os which are possibly related to a fault, it is necessary to distinguish the behavior of single I/Os during the evolution of an observed or modeled I/O vector sequence. The next section introduces appropriate functions which deliver the necessary information.

6.1.2 I/O driven system monitoring

Definition of the I/O behavior

In section 3.1 it has been shown that a closed-loop DES evolution leads to an I/O vector sequence. If the considered system or its model produce a new I/O vector, this is the result of at least one I/O changing its value. If a binary I/O changes its value, an edge can be observed (see figure 6.3):

Definition 40 (Edges). For each controller I/O IO_i there exist three edges: IO_i_0 to indicate a change from 1 to 0 (falling edge), IO_i_1 to indicate a change from 0 to 1 (rising edge) and IO_i_ε to indicate no change in value:

$$E = \{IO_i\_0, IO_i\_1, IO_i\_\varepsilon \quad \forall 1 \leq i \leq m\}$$

with m denoting the number of controller I/Os in the system.

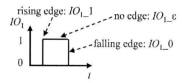

Figure 6.3: Rising edge of an I/O

In order to determine the edges appearing if two arbitrary I/O vectors are considered, an edge function is defined:

Definition 41 (Edge function). Let $IO_i(j)$, $IO_i(k)$ be the i-th controller I/O in the j-th and k-th I/O vector.

$$Edge(IO_i(j), IO_i(k)) = \begin{cases} IO_i\_1 & \text{if } IO_i(j) = 0 \quad \text{and} \quad IO_i(k) = 1 \\ IO_i\_0 & \text{if } IO_i(j) = 1 \quad \text{and} \quad IO_i(k) = 0 \\ IO_i\_\varepsilon & \text{if } IO_i(j) = IO_i(k) \end{cases}$$

delivers the resulting edge of the i-th I/O from the comparison of its values from the controller I/O vectors $u(j)$ and $u(k)$.

Instead of $Edge(IO_i(j), IO_i(k))$ we also write $Edge(u(j)[i], u(k)[i])$, i.e. we address the i-th controller I/O of vector u with $u[i]$. Since more than one I/O can change its value when new I/O vectors are produced, we define the evolution set that summarizes the edges 'between' two I/O vectors:

Definition 42 (Evolution set).

$$ES(u(j), u(k)) = \bigcup_{i=1}^{m} \{Edge(u(j)[i], u(k)[i]) \in E \backslash IO_i\_\varepsilon\}$$

determines the set of rising and falling edges between two I/O vectors $u(j)$ and $u(k)$.

Figure 6.4 shows an example for the evolution set resulting from the comparison of two I/O vectors.

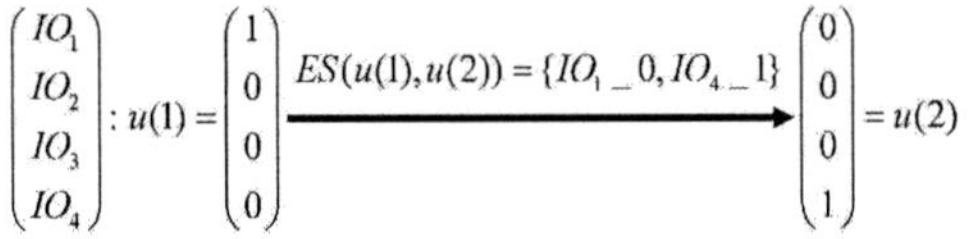

$$\begin{pmatrix} IO_1 \\ IO_2 \\ IO_3 \\ IO_4 \end{pmatrix} : u(1) = \begin{pmatrix} 1 \\ 0 \\ 0 \\ 0 \end{pmatrix} \xrightarrow{ES(u(1),u(2)) = \{IO_1\_0, IO_4\_1\}} \begin{pmatrix} 0 \\ 0 \\ 0 \\ 1 \end{pmatrix} = u(2)$$

Figure 6.4: Example for the evolution set

The NDAAO as part of the evaluator

As explained in section 2.3, the first step in diagnosis is fault detection. In our case, fault detection is performed using the NDAAO as fault-free system model. Figure 6.1 shows that the model is used in the evaluator to reproduce the observed system output.

A more detailed view of this procedure is given in figure 6.5. The part with dashed lines will be introduced in the next sections. The current I/O vector $u(t)$ exhibited by the closed-loop DES is given to the evaluator. Following algorithm 11, the evaluator tries to determine the current state of the NDAAO. The main idea of this algorithm is to determine a current state estimation following the observed evolution set.

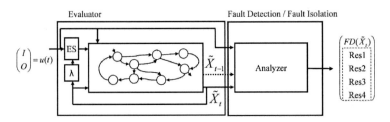

Figure 6.5: Structural scheme for fault detection and isolation with the NDAAO

$\lambda(x)$ determines the output of an NDAAO state according to definition 16 and ES is the evolution set according to definition 42. $f(x)$ contains the set of possible following

---

Algorithm 11 Evaluator algorithm

Require: New observed I/O vector $u(t)$ and former state estimation $\widetilde{X}_{t-1}$
1: **if** $|\widetilde{X}_{t-1}| > 0$: **then**
2:    $\widetilde{X}_t := \{x \in X \,|\, (\exists x_{pre} \in \widetilde{X}_{t-1} \wedge$
$$x \in f(x_{pre}) \wedge ES(\lambda(x_{pre}), \lambda(x)) = ES(\lambda(x_{pre}), u(t)))\}$$
3: **else**
4:    $\widetilde{X}_t := \{x \in X \,|\, \lambda(x) = u(t)\}$
5: **end if**
   $\widetilde{X}_{t-1} := \widetilde{X}_t$
6: **return** $\widetilde{X}_t$

---

states of state x. The algorithm checks if the former state estimation $\widetilde{X}_{t-1}$ contains at least one state. If the observation is being initialized or after a fault has been detected, this set is empty. In case of an empty estimation ($|\widetilde{X}_{t-1}| = 0$), the evaluator determines each NDAAO state with the observed I/O vector as output as potential current state (see line 4 of algorithm 11) and adds it to the current state estimation $\widetilde{X}_t$. If the former state estimation $\widetilde{X}_{t-1}$ was not empty, the algorithm checks the observed evolution set of I/O edges in order to determine the current state estimation. Each NDAAO state that can be reached by the observed edges starting in one of the states from the former state estimate $\widetilde{X}_{t-1}$ is added to the current state estimation $\widetilde{X}_t$. During fault-free system operation the state estimate is unambiguous ($|\widetilde{X}_{t-1}| = 1$) and the next current state of the NDAAO is the successor of the last current state that can be reached by producing the observed evolution.

The resulting state estimation $\widetilde{X}_t$ of the evaluator as well as the observed I/O vector $u(t)$ are given to the analyzer where the fault detection policy and the fault isolation operations are implemented. The fault detection policy is given by

$$FD(\widetilde{X}_t) = \begin{cases} fault & \text{if } |\widetilde{X}_t| \neq 1 \\ OK & \text{if } |\widetilde{X}_t| = 1 \end{cases} \tag{6.1}$$

A fault is detected if the evaluator cannot determine an unambiguous state estimation. Figure 6.6 shows an example of the state estimation and fault detection process. In the lower part of the figure, the NDAAO is depicted. In the part 'observation' an observed I/O vector sequence is shown. Between NDAAO states and between I/O vectors, the resulting edges are given. It is assumed that the algorithm starts without a defined initial NDAAO state x_0 like it is the case if the start of the diagnosis procedure is not synchronized with the start of the closed-loop DES. After the observation of the first I/O vector, the evaluator algorithm goes to line 4 since $|\widetilde{X}_{t-1}| = 0$. The evaluator determines the NDAAO states with $\lambda(x) = u(t)$ and adds them to $\widetilde{X}_t$. State x_1 and state x_3 have the first observed I/O vector as output. The fault detection policy returns 'fault' since the state estimation is not unambiguous. This changes as soon as the next I/O vector is observed which leads to IO_1_0. Starting from x_1 or x_3 only x_2 can be reached by producing the same evolution set. Hence, the state estimation only contains one state (line 2 of algorithm 11) which makes the fault detection policy declare 'OK'.

The state estimation proceeds with the next observed I/O vectors by determining the state trajectory x_3, x_4. When the last I/O vector is observed, the resulting edge IO_1_0 cannot been produced by leaving state x_4. Hence, the state estimation from line 2 in algorithm 11 results in an empty set which leads to fault detection. During the initial phase of the observation, a fault is detected until the state estimation becomes unambiguous. This can be avoided if an initial NDAAO state x_0 is given and the start of the diagnosis process and the closed-loop DES are synchronized.

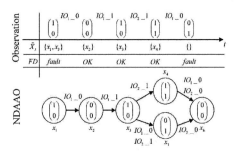

Figure 6.6: Example for state estimation and fault detection

For an NDAAO identified with a given identification tuning parameter k, it is possible to give an upper bound for the number of successively observed fault-free controller I/O vectors to determine an unambiguous state estimate $|\widetilde{X}_t| = 1$:

Theorem 7 (Efficiency of the state estimation algorithm). Given is an NDAAO identified with parameter k according to algorithm 1. If algorithm 11 is used to estimate the current NDAAO state starting with an empty estimate $|\widetilde{X}_t| = \{\}$, the observation of a fault-free I/O vector sequence of length k which is part of the identified language, $w^k \in L^k_{Ident}$ is sufficient to determine an unambiguous state estimate $|\widetilde{X}_t| = 1$.

Proof of theorem 7. If the identification tuning parameter k was chosen such that $L^k_{Ident} \supseteq L^k_{Orig}$ can be stated according to assumption 5, each fault-free I/O vector sequence of lenght k builds a word $w^k \in L^k_{Ident}$. From line 1 of algorithm 1 on page 45 it follows that exactly one state exists with[1] $\lambda(x) = w^k$. Lemma 2 makes sure that this state can only be reached by producing the word w^k. Algorithm 11 determines all state trajectories producing w^k. Since each trajectory must end in the same state, the state estimate is unambiguous after processing w^k. □

This shows that too high a value for the identification tuning parameter k can be a disadvantage if the initial state is not given. Nevertheless, in most practical cases the algorithm will admit a solution in less than k I/O vectors. The worst case only appears if several states exist with the same I/O vector as output. If only one state exists with the first I/O vector of the observed w^k as output, line 4 of algorithm 11 already delivers an unambiguous state estimate.

---

[1] $\widetilde{\lambda}(x)$ represents w^k being the state output in step 1 of the identification algorithm before being replaced by its last letter (see definition 22 on page 48)

Probabilistic evaluation of the NDAAO fault detection capability

After the definition of the fault detection procedure and the according policy it is possible to evaluate the capability of a given NDAAO to detect faults. If no fault is currently detected, the current state estimate is unambiguous ($|\widetilde{X}_{t-1}| = 1$) and thus consists of exactly one current NDAAO state. The principle of the evaluator algorithm 11 is to find a next NDAAO state by analyzing the leaving transitions of the current state. If the observed evolution set can be reproduced by taking one of the transitions of the current state, the according target state becomes the new current state. If a fault leads to an I/O vector producing one of these evolution sets, it is thus not possible to detect it. In the following, an upper bound and an estimate for the probability of accepting an I/O vector induced by a fault are derived.

First, we define three probabilistic events:

- F: Fault leading to an altered behavior of at least one controller I/O in the next I/O vector

- E_i: An I/O vector resulting from i edges occurs

- A: An I/O vector does not lead to fault detection in the evaluator algorithm (it can be reproduced by one of the following states of the current state)

With this definition of F, faults are only considered if they are potentially detectable by analyzing the next occurring I/O vector. Sooner or later most faults fall in this category even if they are not immediately detectable upon their appearance: Either they lead to a change in value of an I/O which is not expected or they prevent an I/O from changing its value although this is expected. Using these basic events, the probability of accepting an I/O vector induced by a fault is given by

$$P(A/F) = \sum_{i=1}^{m} P(A \cap E_i | F) \tag{6.2}$$

with m denoting the number of controller I/Os. $P(A \cap E_i | F)$ is the probability of $A \cap E_i$ under the condition F (Papoulis and Unnikrishna, 2002) which means the probability of accepting an I/O vector resulting from i edges under the condition that there is a fault leading to an altered behavior of at least one controller I/O in the considered I/O vector.

This equation can be rewritten as follows:

$$\sum_{i=1}^{m} P(A \cap E_i | F) = \sum_{i=1}^{m} \frac{P(A \cap E_i | F) P(F) P(E_i \cap F)}{P(F) P(E_i \cap F)} \tag{6.3}$$

$$= \sum_{i=1}^{m} \frac{P(A \cap E_i \cap F) P(E_i | F)}{P(E_i \cap F)} \tag{6.4}$$

$$= \sum_{i=1}^{m} P(A | E_i \cap F) P(E_i | F) \tag{6.5}$$

$$\approx \sum_{i=1}^{m} P(A | E_i) P(E_i | F) \tag{6.6}$$

The last equation holds since $P(A|E_i) \approx P(A|E_i \cap F)$: Usually there is only a very small number of I/O vectors which is valid in a given situation. If a sufficiently large I/O vector is taken, $E_i \approx F \cap E_i$ since most of the I/O vectors resulting from i edges are faulty most of the time.

The two remaining probabilities $P(A|E_i)$ and $P(E_i|F)$ can be estimated. $P(A|E_i)$ denotes the probability of accepting an I/O vector under the condition that it is the result of i edges. To get this probability, the following function is defined:

Definition 43 (Number of transitions with i edges). The function

$$LTrans(x, i) = \sum_{\forall x' \in f(x)} \begin{cases} 1 & \text{if } |ES(\lambda(x), \lambda(x'))| = i \\ 0 & \text{else} \end{cases}$$

delivers the number of following states x' of a given state x which are reached by producing exactly i edges in the according evolution set.

A conservative estimate for $P(A|E_i)$ is the probability of accepting an I/O vector resulting from i edges from the NDAAO state with the *most leaving transitions with exactly i edges*:

$$P(A|E_i) \leq \frac{\max_{\forall x \in X}(LTrans(x, i))}{\binom{m}{i}} \tag{6.7}$$

$\binom{m}{i}$ is the binomial coefficient of the number of controller I/Os m and the number of edges i. The binomial coefficient gives the number of possible I/O vectors resulting from i edges: In a given I/O vector there are $\binom{m}{i}$ possible ways to choose i I/Os and to alter them such that a new I/O vector is created. With $\max_{\forall x \in X}(LTrans(x, i))$, the maximum number of transitions leaving one state in the NDAAO and producing exactly i edges is given. The coefficient gives the probability that an arbitrary I/O vector resulting from i edges is accepted in the NDAAO state with the most transitions with i edges.

It is also possible to replace $P(A|E_i)$ with the following equation instead of using the conservative estimate from equation 6.7:

$$\overline{P}(A|E_i) \approx \frac{\sum_{\forall x \in X} \frac{LTrans(x,i)}{|X|}}{\binom{m}{i}} \tag{6.8}$$

In this equation the *average* number of transitions leaving an NDAAO state and leading to i edges is calculated. This average number is then divided by the number of possible I/O vectors resulting from i edges. If each NDAAO is equally likely to be the current state when a faulty I/O vector occurs, this represents the probability of accepting the I/O vector.

The second remaining probability in equation 6.6 is $P(E_i|F)$. It denotes the probability that a faulty I/O vector is the result from i edges. It can be interpreted as the expected part of faults leading to an I/O vector with i edges. Usually, the precise value for $P(E_i|F)$ can only be roughly estimated. In the following, we assume that $P(E_i|F)$ can be conservatively estimated by an educated guess keeping to the following considerations: It is reasonable to expect a large portion of faults leading to only a few edges and to only have a small portion of faults leading to many edges (e.g. 80% leading to one edge, 10% leading to two edges etc.). This is based on the principle of Parsimony (see page 22) were diagnosis is understood as a conjecture that some minimal set of components is faulty. Following this principle, it can be assumed that a fault typically affects only a small number of components and thus I/Os.

The result of the preceding considerations are an upper bound and an expected average for the probability of accepting an I/O vector induced by a fault:

$$\sum_{i=1}^{m} P_{max}(A \cap E_i | F) \le \sum_{i=1}^{m} \left(\frac{\max_{\forall x \in X}(LTrans(x,i))}{\binom{m}{i}} \cdot P(E_i | F) \right) \tag{6.9}$$

$$\sum_{i=1}^{m} \overline{P}(A \cap E_i | F) \approx \sum_{i=1}^{m} \left(\frac{\sum_{\forall x \in X} \frac{LTrans(x,i)}{|X|}}{\binom{m}{i}} \cdot P(E_i | F) \right) \tag{6.10}$$

where $P(E_i | F)$ has to be determined by an educated guess. For the example in figure 6.7 the two measures are calculated. The according values for $P(E_i | F)$ are defined in table 6.1. As a conservative estimate, we assume that only faults leading to one or two edges can occur. It is obvious that using the automaton in figure 6.7, faults leading to three edges can certainly be detected since it does not have a transition with three edges. Since this portion of faults is also added to the expected portion of faults leading to one or two edges, table 6.1 is a conservative estimate. In systems with many controller I/Os, it is especially conservative to assume a larger expected portion of faults with few edges: The weight of faults with many edges in the two probabilities is relatively low since the binomial coefficient $\binom{m}{i}$ of controller I/Os and edges becomes very large if $1 \ll i \ll m$.

It can be seen that state x_4 has the most leaving transitions with one edge. The result is

$$\frac{\max_{\forall x \in X}(LTrans(x,1))}{\binom{3}{1}} P(E_1 | F) = \frac{3}{3} \cdot 0.8 = 0.8 \tag{6.11}$$

For faults leading to two edges (here, x_1 must be considered) we get

$$\frac{\max_{\forall x \in X}(LTrans(x,2))}{\binom{3}{2}} P(E_2 | F) = \frac{2}{3} \cdot 0.2 = 0.01\overline{3} \tag{6.12}$$

If equation 6.11 and 6.12 are summed up, the result is $P_{max}(A \cap E_i | F) \le 0.81\overline{3}$ indicating that there is an upper bound of about 81% for the probability of accepting an I/O vector induced by a fault.

number of edges i	1	2
$P(E_i | F)$	0.8	0.2

Table 6.1: Definition of $P(E_i | F)$ for the example

125

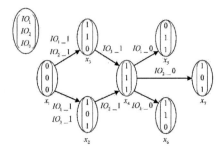

Figure 6.7: Example for the probability measures

The average probability can also be estimated (also assuming the values from table 6.1 for $P(E_i|F)$). The result is

$$\overline{P} \approx \frac{\frac{5}{7}}{3} \cdot 0.8 + \frac{\frac{2}{7}}{3} \cdot 0.2 = 0.21$$

The estimate for the average probability of erroneously accepting a faulty I/O vector is about 21%.

The two measures have a certain similarity with the notion of *diagnosability* from definition 10 on page 16. The notion diagnosability refers to the capability of a model including fault-free *and* faulty behavior to detect a given (and modeled) fault in a finite number of steps. It is thus some kind of guarantee that certain faults can always be detected. This guarantee only holds if a system behaves exactly as captured in the model when the considered fault occurs. The measures introduced in this section are helpful to assess how accurately the fault-free and the faulty behavior can be distinguished by a model. Although they do not guarantee that some given fault can eventually be detected, they give an estimate for the probability of accepting an I/O vector induced by a fault. If this probability is very low, the given model is appropriate for fault detection purposes. Using conservative assumptions for $P(E_i|F)$, the two probabilities are thus helpful to assess the fault detection capability of a given NDAAO. Precise values for the case study system will be given at the end of this chapter.

6.1.3 Residuals as generic fault symptoms

Unexpected behavior

After a fault has been detected by the analyzer using equation 6.1, the next step is to determine which sensor, actuator or hardware part of the plant is possibly affected. Since sensors and actuators are directly connected to controller I/Os, we want to give a small number of I/Os that could be related to the fault. In the following two subsections, four residuals will be introduced that formalize generic fault symptoms. In order to calculate the residuals, the state estimation before a fault was detected is supposed to be unambiguous ($|\widetilde{X}_{t-1}| = 1$). Hence, only one NDAAO state was formerly considered as possible current state. If this condition does not hold, the residuals are not calculated.

The first class of residuals has the aim to isolate faults that led to an observed behavior that was *unexpected* in the given context. The current context is defined by the last estimated current state $\tilde{x}$ in the automaton. The first residual is defined as

$$Res1(\tilde{x}, u(t)) = ES(\lambda(\tilde{x}), u(t)) \setminus \bigcup_{\forall x' \in f(\tilde{x})} ES(\lambda(\tilde{x}), \lambda(x')) \qquad (6.13)$$

With $ES(\lambda(\tilde{x}), u(t))$, the rising and falling edges are determined that are observed when comparing the I/O vector of the last estimated current state and the I/O vector that led to fault detection. This set represents what actually happened when the fault was detected.

$\bigcup_{\forall x' \in f(\tilde{x})} ES(\lambda(\tilde{x}), \lambda(x'))$ represents the union of the sets of rising and falling edges when the last estimated current state and each of its direct successor states $(x' \in f(x))$ are considered. It represents the expected behavior. The set difference of the observed $(ES(\lambda(\tilde{x}), u(t)))$ and the expected $(\bigcup_{\forall x' \in f(\tilde{x})} ES(\lambda(\tilde{x}), \lambda(x')))$ behavior is built in the residual equation to get the *unexpected* I/O behavior. In $Res1$, the expected behavior is given by the union of each possible following behavior of the last estimated current state. A stricter formulation of the expected behavior is used in the second residual:

$$Res2(\tilde{x}, u(t)) = ES(\lambda(\tilde{x}), u(t)) \setminus \bigcap_{\forall x' \in f(\tilde{x})} ES(\lambda(\tilde{x}), \lambda(x')) \qquad (6.14)$$

Instead of a union over the expected behavior of the possible following states, an intersection is used. The intersection delivers the edges that *must* be observed no matter which following state in the model is taken. It is obvious that $Res1 \subseteq Res2$ since $\bigcup_{\forall x' \in f(\tilde{x})} ES(\lambda(\tilde{x}), \lambda(x')) \supseteq \bigcap_{\forall x' \in f(\tilde{x})} ES(\lambda(\tilde{x}), \lambda(x'))$. Figure 6.8 shows an example for unexpected behavior that led to fault detection (instead of the I/O vectors only the edges between two states are given): from the estimated current NDAAO state x_1 it is not possible to take a transition that has exactly the observed falling edges IO_3_0 and IO_4_0. Hence a fault is detected. The result of $Res1$ is:

$$Res1(\tilde{x}, u(t)) = \{IO_3\_0, IO_4\_0\} \setminus (\{IO_1\_0, IO_2\_1, IO_4\_0\} \cup \{IO_1\_0, IO_2\_1\})$$
$$= \{IO_3\_0\}$$

This result means that IO_3_0 was unexpected in the current context. This implies that the system operator should check the sensor or actuator that is connected with IO_3. However, it is possible that the fault cannot be found at this component. If this is the case, $Res2$ should be calculated in order to use a stricter formulation of the expected behavior. In $Res1$ each possible following behavior is subtracted from the observation. Using $Res2$ only the behavior that must occur no matter which regular following behavior is considered:

$$Res2(\tilde{x}, u(t)) = \{IO_3\_0, IO_4\_0\} \setminus (\{IO_1\_0, IO_2\_1, IO_4\_0\} \cap \{IO_1\_0, IO_2\_1\})$$
$$= \{IO_3\_0, IO_4\_0\}$$

This result implies that the occurrence of a change in value of IO_4 is not always expected in the current context. Hence, it is another possible fault candidate.

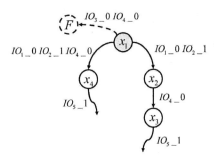

Figure 6.8: Example for an unexpected behavior

Missed behavior

In contrast to an observed but unexpected behavior it is also possible that a faulty component can be isolated by determining a *missed* event. Set operations that help to isolate an expected but unobserved behavior are given by the third and the fourth residual.

$$Res3(\widetilde{x}, u(t)) = \bigcap_{\forall x' \in f(\widetilde{x})} ES(\lambda(\widetilde{x}), \lambda(x')) \backslash ES(\lambda(\widetilde{x}), u(t)) \qquad (6.15)$$

$Res3$ is the set difference of the edges that are expected no matter which following state is taken $(\bigcap_{\forall x' \in f(\widetilde{x})} ES(\lambda(\widetilde{x}), \lambda(x')))$ and the edges that have been observed $(ES(\lambda(\widetilde{x}), u(t)))$. Each rising or falling edge that must occur when the estimated current state is left but has not been observed is part of $Res3$. The expected behavior is represented by the intersection of each possible following behavior. It is also possible to give a less strict formulation of the expected behavior by using the union operation instead of the intersection:

$$Res4(\widetilde{x}, u(t)) = \bigcup_{\forall x' \in f(\widetilde{x})} ES(\lambda(\widetilde{x}), \lambda(x')) \backslash ES(\lambda(\widetilde{x}), u(t)) \qquad (6.16)$$

Since $Res3 \subseteq Res4$, the result of $Res4$ is usually less restrictive than $Res3$, i.e. it contains more elements. The example of figure 6.9 illustrates $Res3$ and $Res4$. The observation of the edge IO_3_0 leads to fault detection from x_1, the estimated current NDAAO state. Applying the residuals results in $Res3 = \{IO_1\_0\}$ and $Res4 = \{IO_1\_0, IO_2\_1\}$. The same procedure as explained for $Res1$ and $Res2$ should be started: First, the component connected with IO_1 should be checked. If the repair team does not find a fault at this component, it should proceed with checking IO_2 which is additionally part of $Res4$.

Especially if industrial closed-loop DES with many controller I/Os are considered, the residuals can help to get a relatively small set of I/Os that could be related to the fault. A maintenance operator can then check the possibly faulty sensors or the related actuators. Actuators are related to a sensor if their activation or deactivation can have an influence on the sensor value like explained for the causal I/O map of definition 39 on page 106. It is also possible to reduce the set of fault candidates by further analyzing

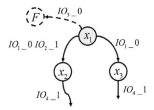

Figure 6.9: Example for a missed behavior

the system behavior following fault detection. A special state estimation algorithm has been proposed for this purpose in (Roth et al., 2009c).

6.1.4 Monolithic fault detection and isolation of the case study system

Treating the case study with the monolithic diagnosis approach necessitates a monolithic NDAAO. As shown in section 3.5, it is possible to identify a complete model with $k = 2$ for the case study system if *a system evolution consists of treating two work pieces*. This scenario is considered in this chapter. It is clear that even for the case study system it is not possible to test the method for each possible fault. The presented examples have been chosen such that they represent typical examples for faulty behavior.

At first, the identified NDAAO is evaluated with the probabilistic measures from section 6.1.2. For the case study, the a priori probability $P(E_i|F)$ denoting the probability that a faulty I/O vector is the result from i edges is defined in table 6.2. It is conservatively assumed that the portion of faults leading to an I/O vector with one edge is 80%. Another conservativism is not to consider faults leading to more than three edges (see page 124 for an explanation on why this is conservative). Calculating the upper bound and the average probability for accepting an I/O vector induced by a fault results in

$$\sum_{i=1}^{m} P_{max}(A \cap E_i|F) \leq \frac{3}{\binom{30}{1}} \cdot P(E_1|F) + \frac{2}{\binom{30}{2}} \cdot P(E_2|F) + \frac{2}{\binom{30}{3}} \cdot P(E_3|F)$$
$$\leq 0.0807$$

$$\sum_{i=1}^{m} \overline{P}(A \cap E_i|F) \approx \frac{0.58}{\binom{30}{1}} \cdot P(E_1|F) + \frac{0.15}{\binom{30}{2}} \cdot P(E_2|F) + \frac{0.31}{\binom{30}{3}} \cdot P(E_3|F)$$
$$\leq 0.0151$$

The calculation shows that in the worst case there is a probability of 8.07% for accepting an I/O vector induced by a fault. If each NDAAO state is expected to be equally likely to be the current automaton state during online monitoring, the probability is only 1.51%. These low values show that the automaton is appropriate for fault detection purposes.

number of edges i	1	2	3
$P(E_i \vert F)$	0.8	0.15	0.05

Table 6.2: Definition of $P(E_i \vert F)$ for the case study

The first example is a fault at the sensor that is connected with the controller input $I3.3$. The fault that has been introduced artificially is a short circuit that causes the sensor to switch its signal from 1 to 0. The fault has been introduced when the first work piece was treated by the second machine (vertical milling). At the same time, the second work piece gets drilled in the first station. The short circuit has been produced when milling with the second of the three tools in station 2 started and the milling head left the top position. This situation is depicted in figure 6.10. After the observation of $I2.2\_1$ indicating that the milling head just left its home position, the evaluator algorithm delivers state x_{18} as single state estimate. Hence, no fault is detected. Then the falling edge $I3.3\_0$ is observed that occurs due to the fault. The evaluator algorithm does not find a following state of x_{18} that can be reached by this edge. Since $\widetilde{X}_t$ is empty, a fault is detected due to equation 6.1. Consequently the residuals are applied with $\widetilde{x} = x_{18}$ as the former unique estimate.

$$Res1(\widetilde{x}, u(t)) = \{I3.3\_0\}$$
$$Res2(\widetilde{x}, u(t)) = \{I3.3\_0\}$$
$$Res3(\widetilde{x}, u(t)) = \{\}$$
$$Res4(\widetilde{x}, u(t)) = \{I1.5\_1, I2.3\_0, O1.3\_1, O1.4\_1, O1.5\_1, O2.2\_1, O2.3\_0\}$$

$Res1$ and $Res2$ both result in the same unexpected edge that is caused by the fault. $Res3$ results in an empty set and $Res4$ shows a union of the legal following behavior of state x_{18}. In this case, $Res3$ and $Res4$ do not isolate the fault. $Res1$ and $Res2$ provide the correct fault candidate.

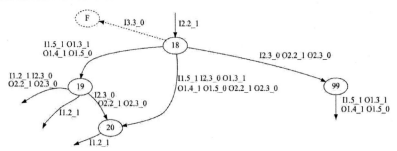

Figure 6.10: Example for a fault at I3.3

The second example is a fault at the sensor that is connected to $I2.4$. This sensor detects if a work piece is at the left most position of station 2 ($I2.4 = 1$). The artificially introduced fault prevents the sensor from switching back to 0 when the work piece has left the leftmost position. The fault was introduced when the first work piece gets transported from station 2 to station 3 and the second work piece from station 1 to station 2. Figure 6.11 shows the described situation. The evaluation algorithm (algorithm 11) determines x_{49} as the actual estimate when the second work piece arrives at position $I2.4$ and produces the rising edge $I2.4\_1$. Now, the fault prevents $I2.4$ from switching again to 0 when the work piece leaves the position due to the start of the conveyor ($O2.5\_1$). The fault is detected when the rising edge of $I3.5$ and its corresponding output changes are observed before $I2.4$ changed back to 0: The sensors detect the arrival of the first work piece in front of machine tool 3 *before* $I2.4$ switched back to 0 which never happened during fault-free system evolutions. The residuals are calculated:

$$Res1(\widetilde{x}, u(t)) = \{I3.5\_1, O3.3\_1, O3.4\_1, O3.5\_0\}$$
$$Res2(\widetilde{x}, u(t)) = \{I3.5\_1, O3.3\_1, O3.4\_1, O3.5\_0\}$$
$$Res3(\widetilde{x}, u(t)) = \{I2.4\_0, O1.5\_0\}$$
$$Res4(\widetilde{x}, u(t)) = \{I2.4\_0, O1.5\_0\}$$

In this case, $Res1$ and $Res2$ do not isolate the fault since they show the observed behavior that is not responsible for the fault. The controller input that is related to the blocked sensor is isolated by both $Res3$ and $Res4$ since it is a missed behavior.

A fault that leads to exactly the same symptoms is a defect at the motor of the conveyor in front of the vertical milling machine ($O2.5$). If the conveyor motor does not start due to the motor fault, the symptoms are comparable: The sensor that is connected to $I2.4$ does not change its value since the conveyor does not transport the work piece away from the left most position of station 2. Hence, if $Res3$ or $Res4$ report a missing change in value at a sensor, it is reasonable to not only check the according sensor but also the actuators that have a direct influence on this sensor

It can be seen that the residuals do not only contain controller inputs but also controller outputs. They are set if a controller input leads to a fulfilled logical condition in the control program. Although outputs reported in the residuals are usually not directly related to the fault, they can be useful to decide whether the fault candidates reported by $Res1$ and $Res2$ or the candidates of $Res3$ and $Res4$ should be analyzed first. If controller inputs appear in $Res1$ or $Res2$, they are supposed to be unexpected. Indeed, the fault-free system model did not expect their change in value at the time of their observation. Nevertheless, if a change in value of an input together with changes in value of controller outputs are observed, it is possible that this was not completely unexpected: The controller program was in an internal state where it waited for the change in value of the input and thus triggered some outputs upon its observation. Given a situation like in the second example, this can help to decide that the observed change in value of $I3.5$ was not unexpected by the controller since it was in an internal state that allowed setting corresponding outputs. In the example scenario it was actually expected that the work piece eventually arrives in front of station 3 which leads to the observation of $I3.5\_1$ and the according output setting to stop the conveyor and start the third machine tool. Hence, the candidates delivered by $Res3$ and $Res4$ should be checked first which delivers the right fault candidate.

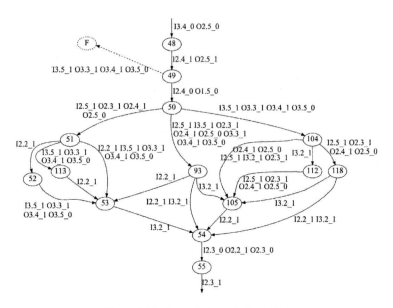

Figure 6.11: Example for a fault at $I2.4$

6.2 Fault Detection and Isolation with the Identified Distributed Models

6.2.1 System monitoring with the distributed model

If the identified distributed models are to be used for online fault detection and isolation, the method introduced in the former section must be adapted. The general idea of using the monolithic and the distributed model is the same: The faults are isolated by determining unexpected and missed behavior. In the distributed model (pyramid structure in figure 6.12) two scenarios for fault detection exist: Firstly, a fault can be detected if the current observation is not reproducible by one of the partial automata. Secondly, the tolerance specification can reach the fault state if the combined behavior of the partial automata exceeds the predefined amount of acceptable unknown behavior. For both cases, appropriate isolation techniques will be proposed in the next section. Figure 6.12 shows the adapted evaluator scheme for the distributed model. It shows the data flow during online monitoring. The precise working principle of the block 'evaluator' is explained in the following. The fault isolation techniques implemented in the block 'analyzer' are given in the next section. It can be seen that the information given to the analyzer are the current state of the tolerance specification, the current state of the POCP and T_{Obs}. T_{Obs} corresponds to a trajectory of I/O vectors observed in case of a fault. It contains each vector observed after the tolerance specification left the OK-state. The required variables are calculated with algorithm 12.

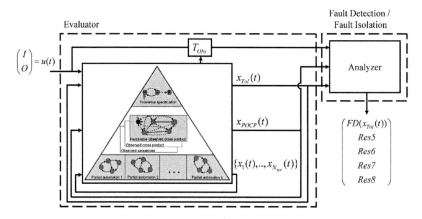

Figure 6.12: Evaluator structure for the distributed model

Algorithm 12 only works if each of the partial automata is in a well defined current state. Hence, it must be started with the set of initial partial NDAAO states. Upon reception of a new I/O vector, the algorithm determines the new automata network situation and derives the current states of the tolerance specification and of the POCP. It is assumed that the tolerance specification has only one state with the fault label F. For the tolerance specification structure, equations 4.8 and 4.9 from page 76 must hold. From line 1 to line 8, the algorithm determines for each partial automaton the current state corresponding to the observed *partial* I/O vector $u_{sys_i}(t)$[2]. The according state must be the former current state itself or one of its successor states (i.e. taken from the set $\{x_i(t-1) \cup f(x_i(t-1))\}$)[3]. If it is not possible to determine a new current state in the i-th partial automaton, the state $x_i(t)$ is not defined and a fault must be detected. In this case, the current state of the tolerance specification is set to the fault state (line 6) to indicate fault detection. In line 9 it is analyzed if the current tolerance specification state represents fault detection. If not, from line 10 to line 14 the next POCP state is determined like in definition 36. In the first case (the condition in line 10 holds) a POCP state exists representing the newly determined states of the partial automata since this combination has been seen during the identification phase. This state becomes the next POCP state. During the construction of the POCP it is made sure that at maximum one POCP state exists for each observed state combination in the automata network (see algorithm 5 on page 81). Each POCP state is connected with all other POCP states. In the second case (line 13), the POCP does not have an according state since the current network situation has not yet been seen. Hence, the joker state with the empty output ε becomes the new current POCP state. In line 15 the next tolerance specification state is determined. Like in definition 36 it must be reached

---

[2]According to definition 26, the partial I/O vector only contains the I/Os of the i-th subsystem

[3]From step 2 of algorithm 1 (the monolithic identification algorithm) it follows that a state in an identified automaton does not have several following states with the same output.

by the same kind of transition as taken in the POCP: If the POCP took an observed transition, the tolerance specification also takes an observed one. If the POCP took an unobserved transition, the tolerance specification also takes an unobserved transition. The functions Θ_{POCP} and Θ_{Tol} deliver the necessary information about observed and unobserved transitions following definition 32.

In line 17 it is analyzed if the tolerance specification just left the OK-state. In this case, the observed sequence T_{Obs} is initialized. It will be used for fault isolation in the next section. If the tolerance specification performed a trajectory from an undecided state to another undecided state or to the fault state (checked in line 20), the new observed I/O vector is added to T_{Obs}. In this case, T_{Obs} had been initialized before. The algorithm returns the current partial automata states, the current states of tolerance specification and of the POCP and the sequence of I/O vectors observed since the tolerance specification left the OK state.

Algorithm 12 Evaluator algorithm for the distributed model

Require: New observed I/O vector $u(t)$, current state combination of the N_{sys} partial NDAAO$_i$ $\{x_1(t-1), \ldots, x_{N_{sys}}(t-1)\}$, current state $x_{POCP}(t-1)$ of the POCP, current state $x_{Tol}(t-1)$ of the tolerance specification

1: **for all** partial NDAAO$_i|1 \leq i \leq N_{sys}$ **do**
2:   **if** $\exists x_i' \in \{x_i(t-1) \cup f(x_i(t-1))\}|\lambda_i(x_i') = u_{sys_i}(t)$ **then**
3:     $x_i(t) := x_i'$
4:   **else**
5:     $x_i(t)$ not defined
6:     $x_{Tol}(t) := x_{Tol}|\lambda_{Tol}(x_{Tol}) = F$
7:   **end if**
8: **end for**
9: **if** $\lambda_{Tol}(x_{Tol}) \neq F$ **then**
10:   **if** $\exists x_{POCP}' \in X_{POCP}|\lambda_{POCP}(x_{POCP}') = \{x_1(t), \ldots, x_{N_{sys}}(t)\}$ **then**
11:     $x_{POCP}(t) := x_{POCP}'$
12:   **else**
13:     $x_{POCP}(t) := x_{POCP}|\lambda_{POCP}(x_{POCP}) = \varepsilon$
14:   **end if**
15:   $x_{Tol}(t) := x_{Tol}'|\big(x_{Tol}' \in f_{Tol}(x_{Tol}(t-1))$
$\wedge \Theta_{POCP}(x_{POCP}(t-1), x_{POCP}(t)) = \Theta_{Tol}(x_{Tol}(t-1), x_{Tol}')\big)$
16: **end if**
17: **if** $\lambda_{Tol}(x_{Tol}(t-1)) = OK \wedge \lambda_{Tol}(x_{Tol}(t)) \neq OK$ **then**
18:   $T_{Obs} := (u(t-1), u(t))$
19: **end if**
20: **if** $\lambda_{Tol}(x_{Tol}(t-1)) \neq OK \wedge \lambda_{Tol}(x_{Tol}(t)) \neq OK$ **then**
21:   add $u(t)$ to the sequence T_{Obs}
22: **end if**
23: **return** $\{x_1(t), \ldots, x_{N_{sys}}(t)\}$, $x_{Tol}(t)$, $x_{POCP}(t)$, T_{Obs}

Algorithm 12 shows that both scenarios leading to fault detection result in the tolerance specification state with the fault label: If a partial NDAAO was not able to reproduce its partial observation, the tolerance specification is set to the fault state in line 6. If the POCP performs an unacceptable number of unobserved transitions, the tolerance specification will finally be led to its fault state in line 15. Hence, the fault detection policy depends on the current state of the tolerance specification only:

$$FD(x_{Tol}(t)) = \begin{cases} fault & \text{if } \lambda_{Tol}(x_{Tol}(t)) = F \\ OK & \text{if } \lambda_{Tol}(x_{Tol}(t)) \neq F \end{cases} \tag{6.17}$$

A fault is detected if the current state of the tolerance specification is the fault state. After a fault has been detected, the model can be reinitialized considering I/O vectors following fault detection[4]: For each partial NDAAO, algorithm 11 from page 121 has to be applied using each new I/O vector until each partial automaton has a unique state estimate. If the fault was detected by a partial NDAAO, this allows reinitializing the according automaton. If the fault was detected by the POCP, this procedure simply determines the new current states in the automata network. After a valid state has been determined for each partial automaton, it can be checked whether a POCP state representing the current network situation exists. If yes, this POCP state becomes the current POCP state and the tolerance specification is reset to the OK state. If no POCP state representing the current network situation exists, it must be continued to use algorithm 11 for each partial automaton until a state combination is reached which is represented by a POCP state. As soon as a POCP state is found, algorithm 12 can restart analyzing the following system behavior with the tolerance specification reset to its OK state.

Analogously to the considerations for the monolithic model, it is also possible to estimate the probability for accepting an I/O vector induced by a fault with the distributed model. The probabilities introduced in section 6.1.2 can be used on the basis of the cross product of the partial automata: The cross product defines the *maximum behavior* which is accepted by the automata network without restriction by the POCP. It can thus be used as a conservative reference for the fault detection capabilities. In section 6.2.3, the according values for the case study system will be given.

6.2.2 Residuals for the distributed model

The fault isolation technique for the distributed model follows the same principle as fault isolation with the monolithic model. The idea is to compare observed and expected I/O behavior after a fault was detected and to determine I/Os with an unexpected or missed behavior. In contrast to the monolithic model, it is not sufficient to compare only the currently observed I/O vector with the expected ones. In the distributed model it is possible that the fault was detected due to an unaccepted combined automata network behavior. Such a behavior makes the tolerance specification leaving its OK state and finally reaching the fault state. Generally, several I/O vectors can have occurred until the tolerance specification reaches its fault state. This sequence contains behaviors

---

[4]This is important for applications where even with the distributed approach it is not possible to completely eliminate false alerts

which caused fault detection. It is thus necessary to analyze the complete observed sequence since the time when the normal behavior (defined by the OK state of the tolerance specification) was left. Algorithm 12 delivers this sequence with T_{Obs}. It represents the *observed* system behavior. Based on

$$T_{Obs} = (u(t-n), u(t-n+1), \ldots, u(t)) \tag{6.18}$$

with n denoting the number of I/O vectors after the tolerance specification left its OK state, it is possible to determine the set of observed edges. They are the union of the observed edges between two successive I/O vectors in the observed trajectory:

$$ES_T(T_{Obs}) = ES(u(t-n), u(t-n+1)) \cup \cdots \cup ES(u(t-1), u(t)) \tag{6.19}$$

with ES from definition 42. Equation 6.19 is an adaptation of the evolution set for analyzing sequences of I/O vectors.

Like in the monolithic case, the observed edges have to be compared with the expected ones. They can be calculated on the basis of expected network trajectories. The expected network trajectories can be derived from the POCP by only taking observed transitions (with $\Theta = true$) since they define fault-free behavior observed during the identification phase. For the analysis of expected edges it is necessary to determine the I/O vectors resulting from a POCP trajectory. The following function determines the I/O vector which is the combined output of a partial automata state combination represented by a POCP state.

$$J_{POCP}(x_{POCP}) = J(\lambda_1(x_1), \lambda_2(x_2), \ldots, \lambda_{N_{sys}}(x_{N_{sys}}))|\{x_1, x_2, \ldots, x_{N_{sys}}\} = \lambda_{POCP}(x_{POCP}) \tag{6.20}$$

It makes use of the join-function from definition 28 on page 68. The partial I/O vectors being the output of the underlying partial automata states are combined. During the construction of the POCP it was made sure that only state combinations leading to a valid I/O vector are represented by a POCP state (see algorithm 4 on page 78).

As shown in equation 6.18, the observed I/O vector sequence consists of n vectors. This means that after the OK state in the tolerance specification was left, n I/O vectors have been observed. To compare the observed and the expected behavior, the POCP trajectories of length n are determined which could have occurred if no fault had been detected. They start in state $x_{POCP}(t-n)$ which was the current POCP state before the tolerance specification left its OK state. The trajectories only take observed transitions because this represents a known fault-free network behavior. In the following equation, each POCP trajectory starting in $x_{POCP}(t-n)$ and only taking observed transitions ($\Theta_{POCP} = true$) is determined. Using the adapted join-function from equation 6.20, the resulting expected I/O vector sequences of length n are determined:

$$\widehat{\Psi}_{Exp}(x_{POCP}(t-n)) = \{J_{POCP}(x_{POCP}(t-n)), \ldots, J_{POCP}(x_{POCP}(t))| $$
$$\Theta_{POCP}(x_{POCP}(j), x_{POCP}(j+1)) = true \ (\forall t-n \le j < t)\} \tag{6.21}$$

Based on T_{Obs} containing the observed behavior and on $\widehat{\Psi}_{Exp}(x_{POCP}(t-n))$ containing the expected behavior, residuals like in the former section are introduced. It is possible to apply the modified evolution set function ES_T from equation 6.19 to each expected I/O vector sequence $\widehat{\psi} \in \widehat{\Psi}$ to determine the set of expected edges.

First, the unexpected behavior is determined.

$$Res5(\widehat{\Psi}_{Exp}(x_{POCP}(t-n)), T_{Obs}) = ES_T(T_{Obs}) \backslash \bigcup_{\forall \widehat{\psi} \in \widehat{\Psi}_{Exp}(x_{POCP}(t-n))} ES_T(\widehat{\psi}) \quad (6.22)$$

$Res5$ takes the observed edges and subtracts the union of the edges occurring in the expected I/O vector sequences. The union contains each edge that would have been occurred in any of the expected behaviors. The resulting edges in $Res5$ were not expected in any known network trajectory and consequently are possible fault locations. Hence, $Res5$ follows the same idea like $Res1$. Like in the case of $Res1$, it is also possible to subtract the intersection of edges in the expected behavior:

$$Res6(\widehat{\Psi}_{Exp}(x_{POCP}(t-n)), T_{Obs}) = ES_T(T_{Obs}) \backslash \bigcap_{\forall \widehat{\psi} \in \widehat{\Psi}_{Exp}(x_{POCP}(t-n))} ES_T(\widehat{\psi}) \quad (6.23)$$

In this case, only the edges expected in each possible network trajectory are subtracted from the observed ones. $Res6$ is thus less strict than $Res5$. If it was not possible to determine the fault among the candidates in $Res5$, it can be reasonable to analyze the additional information of $Res6$.

Using the set operations, it is also possible to determine the *missed* behavior as second generic fault symptom.

$$Res7(\widehat{\Psi}_{Exp}(x_{POCP}(t-n)), T_{Obs}) = \bigcap_{\forall \widehat{\psi} \in \widehat{\Psi}_{Exp}(x_{POCP}(t-n))} ES_T(\widehat{\psi}) \backslash ES_T(T_{Obs}) \quad (6.24)$$

In this case, the intersection of the edges occurring during expected network trajectories is taken. From this set, the observed edges are subtracted. The remaining edges should have occurred in each network trajectory but have not been observed in the I/O vector sequence leading to fault detection. It is also possible to formulate a less strict version of this residual by considering the union of the expected edges:

$$Res8(\widehat{\Psi}_{Exp}(x_{POCP}(t-n)), T_{Obs}) = \bigcup_{\forall \widehat{\psi} \in \widehat{\Psi}_{Exp}(x_{POCP}(t-n))} ES_T(\widehat{\psi}) \backslash ES_T(T_{Obs}) \quad (6.25)$$

$Res8$ usually contains more edges than $Res7$. If the fault cannot be found among the edges in $Res7$, the additional edges in $Res8$ can be taken into consideration.

The following example shows that the residuals can be used in both possible fault detection scenarios: They can be applied if a partial NDAAO was not able to reproduce the observed behavior and if the fault was detected because the combined network behavior exceeded the predefined acceptable amount of yet unknown behavior. First, fault detection by a partial NDAAO is considered.

Figure 6.13 shows the considered scenario. In the example, only the edges occurring between two states are shown. In the POCP, the state output 2A means the combination of state 2 and state A in the partial automata. From the construction of the POCP follows that all POCP states are interconnected. If the transition is not given in the

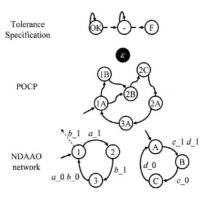

Figure 6.13: Example for fault isolation in the distributed framework

figure, it is unobserved ($\Theta_{POCP} = false$). The first partial NDAAO cannot reproduce the observed edge b_1. As a consequence, algorithm 12 sets the tolerance specification to the fault state in line 6 which leads to fault detection. In line 17 and 18 it is noticed that the tolerance specification just left the OK state. Consequently, the observed trajectory is initialized with the former and the currently observed I/O vector. Applying the modified evolution set function to T_{Obs} leads to:

$$ES_T(T_{Obs}) = \{b\_1\}$$

In the next step it is necessary to determine the expected behavior with $\widetilde{\Psi}(x_{POCP}(t-1))$ according to equation 6.21: Each expected POCP sequence with the length of T_{Obs} which would not have led to fault detection is calculated. Two possible evolutions exist: $1A \rightarrow 1B$ and $1A \rightarrow 2B$. They lead to the edges $\{c\_1, d\_1\}$ for the first evolution and $\{a\_1, c\_1, d\_1\}$ for the second evolution. Applying the residuals leads to:

$$Res5 = \{b\_1\}\backslash(\{c\_1, d\_1\} \cup \{a\_1, c\_1, d\_1\}) = \{b\_1\}$$
$$Res6 = \{b\_1\}\backslash(\{c\_1, d\_1\} \cap \{a\_1, c\_1, d\_1\}) = \{b\_1\}$$
$$Res7 = (\{c\_1, d\_1\} \cap \{a\_1, c\_1, d\_1\})\backslash\{b\_1\} = \{c\_1, d\_1\}$$
$$Res8 = (\{c\_1, d\_1\} \cup \{a\_1, c\_1, d\_1\})\backslash\{b\_1\} = \{a\_1, c\_1, d\_1\}$$

The analysis of $Res5$ and $Res6$ shows that b_1 occurred unexpectedly which allows isolating the fault. $Res7 = \{c\_1, d\_1\}$ shows that two edges have been expected but not observed. $Res8$ additionally contains the edge a_1 because it can occur in one trajectory ($1A \rightarrow 2B$).

In the second example, we assume that an I/O vector sequence was observed leading to the trajectory $1A \rightarrow 2A \rightarrow 2B$ in the POCP. This is only possible by taking two unobserved transitions which leads the tolerance specification to the fault state. The observed edges in this case are $ES_T(T_{Obs}) = \{a\_1, c\_1, d\_1\}$. In the POCP, two possible trajectories starting in 1A with length $|T_{Obs}|$ exists: $1A \rightarrow 1B \rightarrow 2B$ and $1A \rightarrow 2B \rightarrow 2C$. The according expected edges are $\{a\_1, c\_1, d\_1\}$ for the first trajectory and $\{a\_1, c\_1, c\_0, d\_1\}$ for the second trajectory. The residuals result in:

$$Res5 = \{a\_1, c\_1, d\_1\} \backslash (\{a\_1, c\_1, d\_1\} \cup \{a\_1, c\_1, c\_0, d\_1\}) = \{\}$$
$$Res6 = \{a\_1, c\_1, d\_1\} \backslash (\{a\_1, c\_1, d\_1\} \cap \{a\_1, c\_1, c\_0, d\_1\}) = \{\}$$
$$Res7 = (\{a\_1, c\_1, d\_1\} \cap \{a\_1, c\_1, c\_0, d\_1\}) \backslash \{a\_1, c\_1, d\_1\} = \{\}$$
$$Res8 = (\{a\_1, c\_1, d\_1\} \cup \{a\_1, c\_1, c\_0, d\_1\}) \backslash \{a\_1, c\_1, d\_1\} = \{c\_0\}$$

This can be interpreted as follows: None of the observed edges were unexpected since $Res5$ and $Res6$ return empty sets. Since $Res8$ returns $\{c\_0\}$, it can be concluded that c_0 was expected in some regular trajectories but has not been observed. The fault is thus probably related to I/O c which did not change its value.

6.2.3 Distributed fault detection and isolation of the case study system

The fault isolation method for distributed models has been applied to the case study system. The same scenario as in section 4.4 is considered: The system has to treat three work pieces which leads to a high degree of concurrency. One of the main advantages of the method is its capability to compensate for some consequences of inappropriate I/O partitioning. The partitioning process by expert knowledge or by the optimization approach does not always lead to subsystems which allow efficient fault detection and isolation for each I/O. An example for subsystem partitioning with an inappropriately assigned I/O is given in figure 6.14. It can be seen that the I/O $I2.2$ (sensor at the top position of the second machine tool) is assigned to the subsystem with the I/Os from the third machine tool although it is not related to any of its I/Os. Although even worse scenarios are possible, the presented example is significant since it represents a situation where a sensor is not assigned to the same subsystem as its influencing actuators.

Figure 6.14: First scenario for distributed fault isolation

For the partitioning scenario in figure 6.14 a distributed model consisting of three partial automata (each identified with $k = 2$) and the POCP has been identified using the same data base as in section 4.4. The tolerance specification is given in figure 6.15. Differently to the specifications used in section 4.4, it is possible to regain the OK state after some unknown behavior has been observed. If for example *one* unknown transition was taken, it is possible to get back to the OK state after the successive observation of *two* already known transitions in the POCP. This is based on the consideration that the observation of several new transitions is probably an acceptable behavior if the system quickly returns to normal operation. If three unobserved POCP transitions are taken successively (or only interrupted by one or two observed POCP transitions), a fault is detected.

Before a first example is shown, the probability for accepting an I/O vector induced by a fault is calculated for the cross product of the three partial automata using equations 6.9 and 6.10 from page 125. Considering the cross product of the three partial automata refers to using the partial automata without any restriction by a tolerance specification. The probabilities are thus conservative estimates for the fault detection capabilities of the partial automata in conjunction with POCP and tolerance specification. The a priori probability $P(E_i|F)$ denoting the probability that a faulty I/O vector is the result from i edges is defined in table 6.2 on page 130. The resulting values are

$$\sum_{i=1}^{m} P_{max}(A \cap E_i|F) \leq \frac{3}{\binom{30}{1}} \cdot P(E_1|F) + \frac{3}{\binom{30}{2}} \cdot P(E_2|F) + \frac{3}{\binom{30}{3}} \cdot P(E_3|F)$$

$$\leq 0.081 \triangleq 8.1\%$$

$$\sum_{i=1}^{m} \overline{P}(A \cap E_i|F) \approx \frac{1.40}{\binom{30}{1}} \cdot P(E_1|F) + \frac{0.87}{\binom{30}{2}} \cdot P(E_2|F) + \frac{0.94}{\binom{30}{3}} \cdot P(E_3|F)$$

$$\leq 0.037 \triangleq 3.7\%$$

This shows that the distributed model is already appropriate for fault detection purposes even if the network restriction is not considered. A monolithic automaton identified with $k = 1$ on the same data base yields with 5.3% and 2% values which are not significantly better.

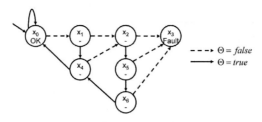

Figure 6.15: Tolerance specification for the case study system

The fault considered in the first scenario is related to $I2.2$. At the beginning of a system evolution, $I2.2$ has the value 0 since the machine tool is at its top position ($I2.2$ is inverted as explained in section 3.5). When the first work piece is processed in the first machine tool, a fault is introduced which makes the input change its value from 0 to 1 without the second machine tool leaving its home position. The partial automaton containing $I2.2$ does not detect a fault because it is waiting for exactly this rising edge. A part of the concerning automaton is shown in figure 6.16. It is the automaton containing $I2.2$ but mainly consisting of I/Os from the third subsystem (third machine tool). During normal system behavior, the partial automaton expects $I2.2\_1$ before any change in value of its remaining I/Os: The first work piece has to be treated in the second machine tool before it is transported to the third one and can influence any of the other I/Os in the automaton. Treating a work piece in the second machine tool leads to a rising edge of $I2.2$ when the milling head is moved down. The partial automaton does not detect that this time the change in value of $I2.2$ occurs too early due to the fault. The fault is finally detected since it leads to an unknown sequence of state combinations in the automata network when the first work piece is treated in the first machine and the partial automaton for the third subsystem is in state x_1. After the POCP performed three unknown transitions, the tolerance specification is led to the fault state and a fault is detected (see algorithm 12).

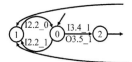

Figure 6.16: Part of the automaton identified for the third subsystem

Applying the residuals for the distributed model leads to the following result:

$$Res5(\widehat{\Psi}_{Exp}(x_{POCP}(t-n)), T_{Obs}) = \{I2.2\_1\}$$
$$Res6(\widehat{\Psi}_{Exp}(x_{POCP}(t-n)), T_{Obs}) = \{I2.2\_1\}$$
$$Res7(\widehat{\Psi}_{Exp}(x_{POCP}(t-n)), T_{Obs}) = \{I1.5\_1, O1.3\_1, O1.4\_1, O1.5\_0\}$$
$$Res8(\widehat{\Psi}_{Exp}(x_{POCP}(t-n)), T_{Obs}) = \{I1.5\_1, O1.3\_1, O1.4\_1, O1.5\_0\}$$

It can be seen that the rising edge at $I2.2$ is part of $Res5$ and $Res6$ which indicates that it occurred unexpectedly. $Res7$ and $Res8$ contain the behavior which was expected but did not occur. It can be seen that exclusively I/Os from the first machine tool are affected. This is due to the fact that usually the first machine tool must finish treating the work piece before a rising edge at $I2.2$ can occur. Since $I2.2$ is part of the first residuals, it is possible to find the fault. Generally, the repair crew must check each I/O contained in one of the residuals. Compared to the total number of 30 I/Os in the system, they still render a relatively sharp fault isolation. The system operator can again use the strategy described in section 6.1.4 to decide which I/O to check first: $Res5$ and $Res6$ only contain $I2.2$ without any change in value of a controller output. Hence, the controller did not expect or react to the change in value of $I2.2$. This is an indicator that $I2.2$ should be checked with priority.

As a second example, we consider a fault leading to a missed I/O behavior. The according I/O partitioning is given in figure 6.17. It is again a wilfully non-perfect solution: In this scenario the I/O belonging to the sensor at the bottom position of the third machine tool ($I3.3$) is assigned to the subsystem exclusively containing the I/Os from the second station.

Figure 6.17: Second scenario for distributed fault isolation

In the second example, the system is working without fault until the first work piece is transported to the third machine. When the work piece arrives in front of the tool, the milling head is moved down to the bottom position ($I3.3$). The sensor changes its value correctly from 1 to 0 (inverted sensor). At this point, a fault is introduced which prevents the input from switching back to 1 when the milling head is moved up after the work piece has been treated. The fault leads to a trajectory consisting of three unobserved POCP transitions which leads the tolerance specification automaton to the fault state. Applying the residuals leads to the following result:

$$Res5(\widehat{\Psi}_{Exp}(x_{POCP}(t-n)), T_{Obs}) = \{\}$$
$$Res6(\widehat{\Psi}_{Exp}(x_{POCP}(t-n)), T_{Obs}) = \{I2.2\_1, O2.2\_0, O2.7\_1, I2.6\_1,$$
$$I1.5\_0, O1.3\_1, O1.4\_1, I1.5\_0\}$$
$$Res7(\widehat{\Psi}_{Exp}(x_{POCP}(t-n)), T_{Obs}) = \{I3.3\_1\}$$
$$Res8(\widehat{\Psi}_{Exp}(x_{POCP}(t-n)), T_{Obs}) = \{I3.3\_1, I1.2\_1\}$$

None of the observed edges was completely unexpected since $Res5$ delivers an empty set. A large set of rising and falling edges have been observed but are not expected in each trajectory of POCP transitions of length $|T_{Obs}|$ not leading to fault detection. The faulty I/O $I3.3$ is part of $Res7$. The rising edge of $I3.3$ is expected in any fault-free POCP trajectory but has not been observed. Since $Res5$ is empty and $Res7$ contains exactly one I/O, it is a reasonable choice to first investigate the state of the sensor connected with I3.3 before any other I/O is considered. The example shows that the missed I/O behavior can be isolated with the distributed models although none of the partial automata detects a fault.

Treating the case study it was possible to show that even faults concerning inappropriately assigned I/Os can be isolated using information of the POCP. Hence, even if the automated partitioning approach does not always lead to 'perfect' results, the upper structure of POCP and tolerance specification allows using the resulting models for fault diagnosis purposes.

7 Industrial Application

7.1 Presentation of the system

In the former chapters the proposed method has been successfully applied to a laboratory case study system. In order to assess the scalability of the method, an industrial production system has been treated. The considered system is a winder used in a production facility producing nonwovens. It is installed in a factory of Freudenberg Vliesstoffe KG in Kaiserslautern.

Freudenberg Vliesstoffe is part of the Freudenberg group. It is a family-owned group of companies developing and producing seals, vibration control technology components, filters, nonwovens, release agents and specialty lubricants as well as mechatronic products. The Freudenberg group employs more than 30000 people in 55 countries and has a balance sheet total of more than 4.6 billion euros (FreudenbergGroup, 2009). The factory located in Kaiserslautern is partially dedicated to nonwovens production. Nonwovens are a special fabric which is made from long fibers, bonded together by chemical, mechanical or heat treatment. Typical fields of use are industrial or automotive filters, medical applications like adhesive plaster or bandages and personal hygiene like diapers. A schematic view of the nonwovens production process is given in figure 7.1. The nonwoven fabric is usually built on the basis of some granulate which in a first step is melted in an extruding process. The melted material is transported to so called spinning jets which stretch and mix single filaments to a web structure. The criss-crossed fibres are welded together by passing through hot rotating cylinders (the calender). The resulting nonwoven fabric is transported to a winder where it is winded on a coil and where the coils are handled.

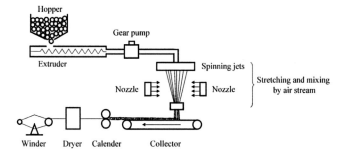

Figure 7.1: Schematic view of a nonwovens production line

The system considered in this chapter is the winding machine at the end of the production process depicted in figure 7.2. The machine winds the fabric on a coil until the desired length is reached. At this point, the incoming fabric is cut and the full coil is taken to a lifting post where it gets manually wrapped. An important feature of the winding machine is its storage: When the coil is wrapped on the lifting post, it is not necessary to stop the remaining production process. The incoming fabric is stored in a storage until an empty coil is installed in the winder. The whole winder is controlled by a Siemens 135U PLC with 336 digital inputs and 256 digital outputs. The PLC has been equipped with an appropriate communication processor such that the same data link as described in section 3.5.2 for the case study system can be used.

Figure 7.2: Winder of the nonwovens production line

Since only with a properly working winder it is possible to continuously produce nonwoven fabric, there is a considerable interest in reducing the necessary time to detect and locate faults in order to restore a functioning production as fast as possible after some fault occurred. Using model-based fault detection techniques based on a manually built model is not possible due to two important obstacles. The first one is the complexity and size of the plant (among other components, the winder contains 20 drives and 38 cylinders of different types). The second one is that the controller program is not available in a formalized form which would allow translating it in a finite state machine and applying e.g. the method of (Philippot et al., 2007). Motivated by these obstacles, the identification-based method presented in this work was chosen to build a fault detection and isolation system.

7.2 Software implementation

To facilitate the installation of the identification based diagnosis system, different software tools have been developed. Figure 7.3 shows the structure of the software. It can be seen that the data collection and the diagnosis tools are run online which leads to realtime constraints. Procedures related to the identification process can be run offline and are thus less time critical. More details on the software used online will be given in section 7.5.

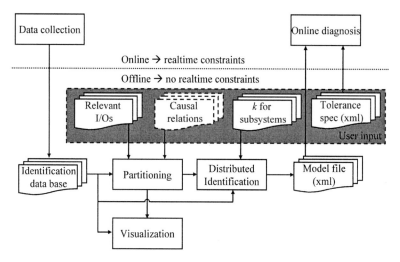

Figure 7.3: Structure of the identification software

The software scheme shows where the user has to generate different inputs. Since not all controller I/Os are relevant for diagnosis, the interesting I/Os must be defined. An example for I/Os which can be ignored are inputs delivering the length of the winded fabric in a binary format or controller outputs connected to lamps or displays. If available, it is also possible to define the causal actuator sensor relations to improve the results of the partitioning approach. When the system has been divided into subsystems, it is necessary to choose appropriate values for the identification parameter k for each subsystem. If the distributed diagnosis procedure is to be applied, the tolerance specification must also be defined.

The software allows visualizing the evolution of the observed system language and the evolution of the optimization algorithm. System partitioning and distributed identification result in a model file in form of an xml-file containing the identified partial automata as well as the POCP. All steps explained in the following sections have been carried out using this software scheme.

7.3 Data analysis

Before the data can be captured, it is necessary to determine controller I/Os which are not relevant for diagnosis. A first analysis resulted in a set of 134 relevant controller I/Os. Figure 7.4 shows the evolution of the observed language L_{Obs}^n for $1 \leq n \leq 3$. It can be seen that even after the observation of 562 system evolutions, L_{Obs}^2 does not converge. Hence, it is not possible to state $L_{Obs}^n = L_{Orig}^n$ for any $n \geq 2$ as required in assumption 5 on page 54 to allow the identification of a monolithic automaton with the minimal value $k = 1$. Two possibilities exist to deal with this situation: The first one is to increase the system observation time and to record more evolutions. Since collecting 562 evolutions already took more than one week, this approach was not possible. Hence, the second possibility was chosen: Dividing the system into subsystems and working with the distributed model.

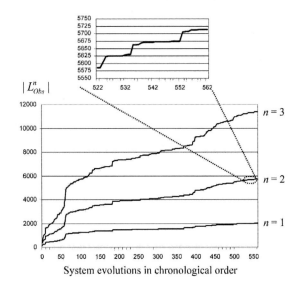

Figure 7.4: Evolution of the observed language of the complete winder

A first partitioning for the considered system was possible by treating the lifting station separately. With the help of system experts it could be determined that this station (with 20 controller I/Os) works almost independently from the remaining winding process. The evolution of the observed language up to length three is shown in figure 7.5 on the left. It can be seen that the language of length three of the lifting station converges and thus allows the identification of an NDAAO based on the observed data with $k = 2$ since $L_{Obs}^{k+1} \approx L_{Orig}^{k+1}$ can be stated. A higher value for k seems not to be reasonable since L_{Obs}^4 does not converge as good as L_{Obs}^3. Additionally, it was possible to roughly analyze the functioning of the lifting station: The most ambiguous situation in terms of distinguishing non-equivalent system states with the same output is comparable to the conveyor example from section 3.4.1. Hence, it could be decided that $k = 2$ is a good compromise between model accuracy and model completeness (to avoid false alerts). The resulting NDAAO has 55 states and 86 transitions. Using a 1.8 GHz processor with 2 GB RAM, the identification could be carried out in 33 seconds. Since the language of the remaining system does not converge for any n (see figure 7.5 on the right), further partitioning is required. Since only few expert knowledge was

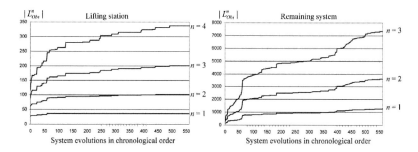

Figure 7.5: Evolution of the observed languages for the lifting station and the remaining system

available, the optimization approach from chapter 5 was chosen to divide the remaining systems into subsystems.

7.4 Distributed identification

Although only limited expert knowledge was available to divide the remaining system into subsystems, it was possible to partially determine causal actuator sensor relations (see section 5.1) with acceptable effort by analyzing the documentation of the machine and by consulting the knowledge of the system operators. The remaining system without the lifting station contains 41 controller outputs and 73 controller inputs. For 26 of the outputs it was possible to determine at least one causally influenced controller input. From the 73 controller inputs, 35 could be determined which are causally influenced by at least one of the controller outputs. For the remaining I/Os it was not possible to get any information concerning causal relations. It was thus possible to partially define the causal I/O map from definition 39 on page 106. With this knowledge, the data-based partitioning approach using the knowledge of causal actuator sensor relations from section 5.3.5 was applied to the system.

For the winding process, the optimal number of subsystems could not be determined using apriori knowledge. Hence, the optimization was first carried out for two subsystems. As optimization criterion, $\widetilde{J}_1(y)$ from equation 5.7 on page 99 (language growth) was taken. Words of length $n = 2$ have been considered. The optimization was parameterized with an initial temperature of $T_0 = 1000$ and the minimal temperature $T_{min} = 8.04 \times 10^{-14}$ leading to 3000 optimization runs with a cooling rate of $CR = 0.99$. The solution difference defining how many I/Os have to change their subsystem from one optimization cycle to another has been chosen to $sD = 3$. Since for many controller outputs the according causal influences could be determined, $sD = 3$ often leads to slightly more than three I/Os changing their subsystem (see section 5.3.5). Using a PC equipped with a 2.83GHz Intel®Core^{TM}2 Quad Core CPU with 3.25 GB RAM, the partitioning took 30 hours. Figure 7.6 shows the evolution of the observed languages for the resulting subsystems. It can be seen that the languages show a better convergence than the languages depicted in figure 7.5 on the right. In order to compare the curves, it is helpful to build the gradient for the language L_{Obs}^n. The gradient for the observed language of length n in the last 100 observed system evolutions is built using the following equation:

$$\Delta_L = \frac{|L_{Obs}^{n,562}| - |L_{Obs}^{n,462}|}{100} \tag{7.1}$$

with $L_{Obs}^{n,462}$ denoting the observed language up to the 462-th system evolution according to definition 37. The gradient is built for the last 100 evolutions since it was decided that for the given application, 100 evolutions are a significant number (see assumption 5 on page 54) to reasonably assess the completeness of the observation.

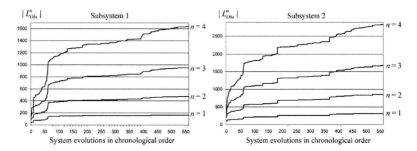

Figure 7.6: Evolution of the observed language for two automatically determined subsystems

The according values for Δ_L of L_{Obs}^2 can be seen in table 7.1. During the last 100 observed system evolutions, the complete system language (without lifting station) produced 2.42 new words of length two on average in each evolution. If this trend is prolonged in the future, it can be expected that a monolithic system model would lead to two to three false alerts in each evolution on average. If the resulting subsystems are considered, it can be seen that the according value of Δ_L is decreased to 0.07 and 0.25 respectively. It can thus be expected that using partial automata identified for the subsystems significantly reduces the number of false alerts since the expected number of non reproducible words is significantly reduced.

Scenario	complete system	subsystem 1	subsystem 2
Δ_L	2.42	0.07	0.25

Table 7.1: Δ_L for the complete system language of length two (without lifting station) and two automatically determined subsystems

If the number of expected false alerts is to be further decreased, it is possible to continue system partitioning. Since concurrency in the two subsystems has already been reduced, it is a straight forward approach to take the two determined subsystems as basis for the next partitioning. Each of the two subsystems was given to the partitioning algorithm which was parameterized like in the former case. Since the subsystems are smaller than the complete system, the number of optimization cycles could be reduced to 1000 (with $T_{min} = 4.3 \times 10^{-5}$). The optimization approach to split each subsystem in two new subsystems took 7 hours. The languages of the resulting four subsystems (two for each of the former subsystems) is shown in figure 7.7.

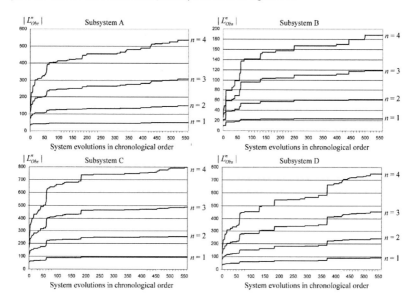

Figure 7.7: Evolution of the observed language for four automatically determined subsystems

The values for the gradient of L^2_{Obs} of the four subsystems are given in table 7.2. It can be seen that dividing subsystem 1 into subsystems A and B yields $\Delta_L = 0.06$ and $\Delta_L = 0$. Since the gradient of subsystem A is not significantly smaller than the gradient of subsystem 1, it can be concluded that dividing subsystem 1 is not reasonable. Indeed, the expected number of non-reproducible words of length 2 for subsystem 1 with 0.06 for each evolution on average (six new words in 100 evolutions) is already acceptably low.

If subsystem C and D are considered, it can be seen that the gradient of subsystem C is almost zero. Since the gradient of subsystem D is only 0.12 (compared to 0.25 for subsystem 2), it can be concluded that dividing subsystem 2 is a reasonable choice since this reduces the number of expected false alerts significantly (almost 50%). In remark 3 on 84 it has been explained that the number of subsystems should be kept as small as possible to sharply distinguish fault-free and faulty behavior. Hence, it was decided that subsystem C and D are not to be further divided. The resulting automata network thus consists of subsystem 1, subsystem C and subsystem D.

Considering the data captured for the industrial application, it can be seen that the convergence of the observed language is not as good as it is in the case study

Scenario	subsystem A	subsystem B	subsystem C	subsystem D
Δ_L	0.06	0.0	0.02	0.12

Table 7.2: Δ_L for the partial system language of length two of four automatically determined subsystems

system treated so far. The main reason is that for the industrial application not each precondition perfectly holds. In chapter 3 it was explained that the considered class of systems is supposed to be an autonomous closed-loop of controller and plant. However, the winder can not be considered as completely autonomous since some of its procedures are triggered by the personnel of the facility using various buttons. Since not each operating crew is following exactly the same procedures, this can be considered as a source of disturbance which makes it hard to observe the complete system behavior. Nevertheless, it can be seen that after having divided the system into subsystems, the observed language of each subsystem rather clearly converges to a stable level.

On the basis of the four subsystems, partial automata have been identified. Except of the lifting station with $k = 2$, the partial identification was carried out with $k = 1$ since only L^2_{Obs} converged good enough to state $L^{k+1}_{Obs} \approx L^{k+1}_{Orig}$ for each subsystem such that theorem 5 and theorem 6 hold. Table 7.3 shows the number of states and transitions for each partial automaton. It also contains information about a monolithic automaton identified for the whole system (with $k = 1$).

	lifting station	subsystem 1	subsystem C	subsystem D	monolith
$|X|$	55	165	94	91	2008
$|f(X)|$	86	307	160	150	3705

Table 7.3: Number of states and transitions for the winder

On the basis of the four partial automata, the POCP has been identified. The resulting automaton has 2167 states. The construction took 14 minutes on a PC with 2.83GHz Intel®Core™2 Quad Core CPU with 3.25 GB RAM.

To assess the fault detection capability of the identified automata network, the probability for accepting an I/O vector induced by a fault is calculated for the cross product of the partial automata using equations 6.9 and 6.10 from page 125. The a priori probability $P(E_i|F)$ denoting the probability that a faulty I/O vector is the result from i edges is defined in table 7.4 by a conservative estimate. The resulting values are

$$\sum_{i=1}^{m} P_{max}(A \cap E_i|F) \leq 0.114 \triangleq 11.4\% \tag{7.2}$$

$$\sum_{i=1}^{m} \overline{P}(A \cap E_i|F) \approx 0.034 \triangleq 3.4\% \tag{7.3}$$

This shows that the distributed model is already relatively well appropriate for fault detection purposes even without restricting the network behavior. A monolithic automaton identified with $k = 1$ yields with 2.98% and 0.6% better values but leads to a significantly larger number of false alerts.

number of edges i	1	2	3
$P(E_i|F)$	0.9	0.05	0.05

Table 7.4: Assertion of $P(E_i|F)$

7.5 Using the model online

An online diagnosis system using the identified models has been installed at the Freuden-berg production facility. The diagnosis program exploiting the models is running on a PC which is also hosting a data base and a web server. The results of the diagnosis algorithm are written to the data base were they can be accessed via the web server. The web server also allows integrating the diagnosis system into the existing SCADA (supervisory control and data acquisition) system. Figure 7.8 shows the principle.

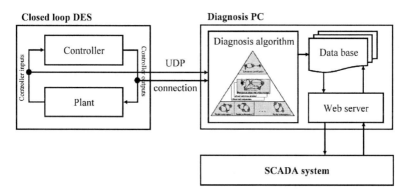

Figure 7.8: Online diagnosis setup

To use the identified models for online diagnosis, the tolerance specification must be designed. It was decided that it is reasonable to accept two successive unknown combined network trajectories as fault-free. After the observation of a third unknown trajectory, a fault should be detected. If after less than two unknown trajectories normal behavior follows, the specification should regain the OK state. To follow this requirement, the tolerance specification from figure 7.9 has been chosen. Observing 200 fault-free system evolutions with the partial automata, the identified POCP and this tolerance specification led to 41 evolutions with a false alert. After each fault, the automata network, the POCP and the tolerance specification have been reinitialized as described in section 6.2.1. This showed that each evolution ended without a second false alert. To compare the distributed and the monolithic approach, the monolithic automaton has also been used to observe the 200 system evolutions. The result are 91 evolutions with at least one false alert. The total number of false alerts was 216. This shows that the distributed approach significantly reduces the number of false alerts.

To evaluate if the number of false alerts can be further reduced, the less restrictive tolerance specification from figure 7.10 was considered. It accepts an unknown network

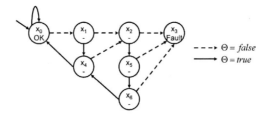

Figure 7.9: First tolerance specification for the industrial application

trajectory of length three before a fault is detected. Observing the same 200 fault-free system evolutions results in 36 false alerts. It is thus not possible to significantly reduce the number of false alerts with the less restrictive specification.

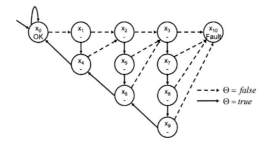

Figure 7.10: Second tolerance specification for the industrial application

Table 7.5 shows the number of false alerts using different models. Using the tolerance specification 1 from figure 7.9 reduces the number of false alerts significantly when compared with the monolithic model. Since false alerts are not totally removed, the distributed model is still sensitive to unknown behavior.

Model	evolutions with false alerts	total number of false alerts
Monolith	91 (45.5%)	216
Distr. + tol. spec. 1	41 (20.5%)	41
Distr. + tol. spec. 2	36 (18%)	36

Table 7.5: False alerts observing 200 fault-free system evolutions

In the industrial application, it was not possible to artificially introduce faults due to the high risk of causing damage in the system. Although the diagnosis system is currently running at the production site of Freudenberg, no real faults occurred until the end of the evaluation time. Nevertheless, the low values for the probability of accepting an I/O vector induced by a fault allows being confident about the fault detection capability of the model.

8 Conclusion

8.1 Summary

The aim of the presented work was the development of a generic model-based diagnosis method which can be applied to the class of closed-loop discrete event systems. Reviewing existing approaches known from literature revealed that two main principles for model-based diagnosis exist: The first one is to use models containing the fault-free as well as the faulty system behavior. In this context, the models must usually be built manually and can only handle faults predefined during the modeling phase. The second possibility for model-based diagnosis is to use models of the fault-free system behavior only. The diagnosis principle is to detect a fault as soon as the current observation cannot be reproduced by the fault-free system model. It has been shown that using fault-free system models has a major advantage for the application to large systems: Since they only contain fault-free behavior, it is possible to construct them using identification methods working on data captured during fault-free system evolutions. Since the considered class of industrial closed-loop DES is usually hard to model manually, the second approach has been chosen in this work.

The first contribution of the thesis is thus in the field of identification. An existing monolithic identification algorithm for the considered class of systems has been improved and a new completeness property has been shown. For the choice of the identification tuning parameter k, new results have been reported. It has been explained how the identification parameter helps to avoid representing several non-equivalent systems states by only one automaton state. Although the monolithic identification approach is capable of delivering very accurate models, it has some shortcomings when systems with a high degree of concurrency are to be treated. For these systems it is often not possible to observe the complete fault-free system behavior in a reasonably short time. The result are many non-reproducible fault-free behaviors leading to false alerts when using the monolithic model for fault diagnosis. To overcome this problem, a new distributed identification method has been developed. The idea is to divide the system into subsystems and to systematically accept a certain amount of unknown combined subsystem behavior. The approach is based on the heuristic that a certain amount of *unknown global* behavior resulting from a combination of regular subsystem evolutions can often be accepted as fault-free because it is similar to the known fault-free behavior. The acceptable amount of new behavior can be given by a tolerance specification automaton. The key of the distributed approach is the division of a given system into appropriate subsystems. A method to *automatically* perform this partitioning based on observed system behavior only and a minimum of system knowledge has been proposed. It uses an optimization technique to solve the combinatorial problem of assigning controller I/Os to subsystems. Since the aim is to get subsystems with weak internal concurrency, optimization criteria to estimate the concurrency in a given subsystem have been introduced.

For the use of the identified models for fault detection and isolation, a method inspired by the concept of residuals known from continuous systems has been proposed. It derives parts of the controller I/O vector which are possibly related to a detected fault following the principle of Parsimony introduced by (Reiter, 1987). For both, the monolithic and the distributed model, residuals have been defined. Using set operations on the observed and expected I/O behavior, the residuals are able to determine unexpected and missed behaviors. The formulation of residuals for missed and unexpected behavior can be understood as the formalization of fault symptoms. Two probabilistic measures to assess the fault detection capability of a given model have been given.

The methods introduced in this work have been applied to a laboratory case study system. The results show that both identification of appropriate models as well as fault detection and isolation of various tested faults are possible. The scalability of the developed approach has been shown with an industrial application. It has been shown that the distributed models significantly reduce the number of false alerts compared to the use of the monolithic model.

8.2 Outlook

For future work several interesting directions exist. The most interesting question is if it is possible to integrate the timed behavior in the proposed method. This would allow detecting and isolating faults leading to a deviant timed behavior and deadlocks. Since the described approach is identification based, a promising way could be to use statistical methods to derive minimum and maximum state durations for the identified models. If the timed behavior can be determined on the subsystem level, it is an interesting question to analyze how the timed subsystem behavior restricts the automata network behavior. It is possible that it restricts the network behavior such that the restriction using the methods from chapter 4 can be relaxed.

Another interesting question is online updating of the system model. If during online diagnosis a non-reproducible behavior has been identified as fault-free by the system operator, this behavior should be included into the model. For the monolithic model, an appropriate online updating algorithm has been proposed in (Roth et al., 2009b). Since in most practical cases the distributed model will be used, the development of an appropriate adaptation for the partial models and the POCP is an interesting question.

The identification procedures introduced in this work have been explicitly designed to deliver appropriate models for fault detection purposes. Since the models describe the functioning of the system under normal conditions, other model-based techniques like re-engineering or validation and verification may also be possible with the identified models. In this context it is an interesting question to find formal requirements of other model-based techniques and to check if the identification methods from this work deliver appropriate models.

9 Extended summaries in German and French

9.1 Kurzfassung in deutscher Sprache

Die Wettbewerbsfähigkeit von Industrieunternehmen hängt maßgeblich von der Produktivität eingesetzter Betriebsmittel wie Anlagen und Maschinen ab. Ungeplante, durch Fehler verursachte Stillstandszeiten müssen so kurz wie möglich gehalten werden. Dies ist nur dann möglich, wenn die Ursachen für die Betriebsstörungen schnell gefunden werden können. Eine Teildisziplin der Automatisierungstechnik befasst sich daher mit der Frage, wie Fehler in technischen Systemen schnell detektiert und möglichst genau lokalisiert werden können.

Zahlreiche industrielle Prozesse lassen sich dabei als ereignisdiskrete Systeme auffassen, die aus einem geschlossenen Kreis von Steuerung und Steuerstrecke bestehen (siehe Bild 9.1). Das Verhalten solcher closed-loop Systeme kann durch Betrachtung der zwischen Steuerung und Strecke ausgetauschten Sensor- und Stellsignale analysiert werden. Ziel der Arbeit ist es, für diese Klasse von Systemen einen generischen Diagnoseansatz zu entwickeln.

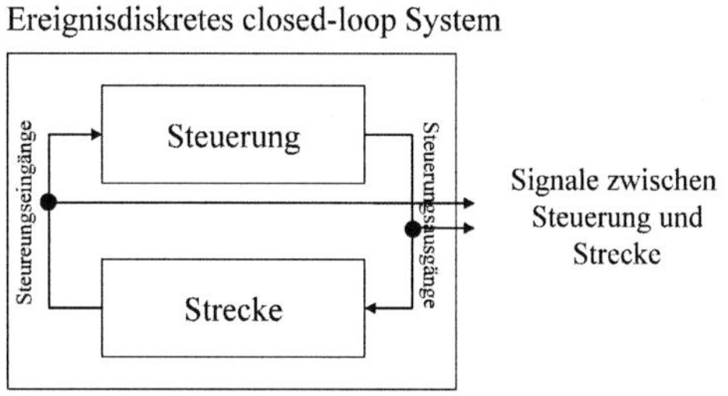

Ereignisdiskretes closed-loop System

Abbildung 9.1: Ereignisdiskretes closed-loop System

Die meisten bekannten Diagnosemethoden für technische Systeme lassen sich in die Kategorien datenbasierte Ansätze, Expertensysteme oder modellbasierte Ansätze einordnen. Eine Analyse der drei Kategorien zeigt, dass modellbasierte Verfahren das größte Potential für den Einsatz in ereignisdiskreten closed-loop Systemen haben: Bei modellbasierten Verfahren wird das aktuell beobachtete Systemverhalten mit dem Verhalten eines Referenzmodells verglichen (siehe Bild 9.2). Auf Basis des Vergleichs kann entschieden werden, ob die aktuelle Situation im System durch einen Fehler hervor gerufen wurde oder Ausdrck normalen Verhaltens ist. Modellbasierte Methoden sind für die betrachtete Systemklasse besonders gut geeignet, da mit Hilfe eines Modells die Systemdynamik gut abgebildet werden kann. Ereignisdiskrete Modelle erlauben es insbesondere den Zustand eines Systems zu modellieren. Ein bestimmtes Systemverhalten kann abhängig vom Systemzustand sowohl Folge normaler Funktion als auch von Fehlfunktion sein. Durch Zustandsmodelle ist eine präzise Beurteilung des beobachteten Verhaltens möglich, als beispielsweise bei rein datenbasierten Verfahren.

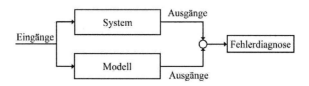

Abbildung 9.2: Prinzip modellbasierter Diagnose

Aus der Literatur bekannte modellbasierte Diagnoseverfahren lassen sich in Methoden mit Fehlermodellen und in Methoden ohne Fehlermodelle unterteilen. Bei Methoden mit Fehlermodellen wird das beobachtete Verhalten mit dem modellierten Fehlerverhalten verglichen. Ist das aktuell beobachtete Verhalten mit einem der modellierten Fehlerverhalten konsistent, wird der betreffende Fehler diagnostiziert. Der Schwachpunkt dieser Ansätze ist, dass nur Fehler, deren Auswirkungen explizit modelliert wurden, erkannt und diagnostiziert werden können. Methoden, die ohne Fehlermodelle arbeiten, umgehen diesen Nachteil. Bei dieser Klasse von Methoden wird ein Fehler immer dann detektiert, wenn sich ein beobachtetes Verhalten nicht mit dem fehlerfreien Modell reproduzieren lässt. Da die Modelle prinzipiell weniger Wissen enthalten als Modelle mit explizit erfasstem Fehlerverhalten, ist das Finden der Fehlerursache oft schwerer als bei Methoden mit Fehlermodellen.

Die zentrale Herausforderung bei der Anwendung modellbasierter Verfahren ist die Modellgewinnung. Bei Systemen im industriellen Maßstab ist das manuelle Modellieren aufwändig und teuer. Daher wurde in dieser Arbeit der Ansatz der Modellidentifikation verfolgt. Auf Basis von Systemdaten, die während fehlerfreier Systemevolutionen aufgezeichnet wurden, wurde zunächst ein Modell in Form eines monolithischen Automaten identifiziert. Der nicht-deterministische autonome Automat mit Ausgabefunktion funktioniert genau wie die betrachtete Klasse ereignisdiskreter closed-loop Systeme als Ereignisgenerator. Ein bereits vorhandener Identifikationsalgorithmus von (Klein, 2005) wurde in dieser Arbeit weiterentwickelt. Der Algorithmus arbeitet auf Basis von Systemverhalten, das während fehlerfreier Systemevolutionen beobachtet wurde. Mit Hilfe eines Tuningparameters $k \in \mathbb{N}^+$ kann die Modellgenauigkeit justiert werden. Ein mit dem Algorithmus identifizierter Automat ist $k + 1$-vollständig. Dies bedeutet, dass die Sprache[1] des Automaten L_{Ident}^{k+1} exakt gleich der beobachteten Sprache der Länge $k + 1$ ist, $L_{Ident}^{k+1} = L_{Obs}^{k+1}$. Führt ein Fehler während der Online-Diagnose mit dem identifizierten Automaten zu einer Sequenz $w \notin L_{Obs}^{k+1} = L_{Ident}^{k+1}$, ist somit garantiert, dass

---

[1] Eine Sprache L^n besteht aus Folgen bis zur Länge n von Ausgangssymbolen (hier Vektoren aus Ein- und Ausgangssignalen der Steuerung)

der Automat diese Sequenz nicht reproduzieren kann. Es ist daher gesichert, dass der Fehler detektiert wird.

Eine wichtige Voraussetzung bei der Wahl eines geeigneten Wertes für k ist, dass die beobachtete Sprache L_{Obs}^{k+1} der fehlerfreien Sprache des Systems L_{Orig}^{k+1} entspricht: $L_{Obs}^{k+1} = L_{Orig}^{k+1}$. Ist dies nicht der Fall, tritt während der Online-Diagnose immer dann ein Fehlalarm auf, wenn eine bisher unbekannte fehlerfreie Sequenz $w \in L_{Orig}^{k+1}$ beobachtet wird, die mit dem Automaten nicht reproduziert werden kann, da $w \notin L_{Obs}^{k+1} = L_{Ident}^{k+1}$ gilt. Als Kriterium für $L_{Obs}^{k+1} \approx L_{Orig}^{k+1}$ kann die Konvergenz von L_{Obs}^{k+1} wie in Bild 9.3 angesehen werden: das System zeigt während der Lernphase mit wachsender Beobachtungszeit immer größere Teile des möglichen fehlerfreien Verhaltens, das in das zu identifizierende Modell überführt wird. Wenn das gesamte mögliche fehlerfreie Verhalten während der Lernphase aufgetreten ist und damit in die beobachtete Sprache überführt wurde, ist zu erwarten, dass diese nicht weiter wächst und gegen einen bestimmten Wert konvergiert.

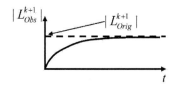

Abbildung 9.3: Konvergenz der beobachteten Sprache

Bei Systemen mit ausgeprägter Nebenläufigkeit kann es sehr lange dauern, bis eine Konvergenz der beobachteten Sprache festgestellt werden kann. Der Grund hierfür ist, dass es für nebenläufige Prozesse sehr viele verschiedene kombinierte Verhalten gibt, die während der Beobachtungsphase erfasst werden müssen. Die hohe Anzahl an Fehlalarmen, die aus dem unvollständig beobachteten Verhalten resultieren, kann schnell dazu führen, dass das Diagnosesystem unbrauchbar wird. Aus wirtschaftlichen Gründen ist es für viele Systeme nicht möglich den Beobachtungshorizont ausreichend zu erweitern bis Konvergenz der beobachteten Sprache eintritt. Um diesem Problem zu begegnen, wurde in dieser Arbeit ein verteilter Identifikationsansatz entwickelt. Kern des Ansatzes ist das Unterteilen des betrachteten Systems in Teilsysteme mit schwacher innerer Nebenläufigkeit. Sind die Teilsysteme geeignet gewählt, konvergieren die beobachteten Teilsystemsprachen sehr viel schneller als dies für die globale Systemsprache der Fall ist. Eine Möglichkeit eine geeignete Systemaufteilung zu ermitteln ist vorhandenes Systemwissen zu nutzen. Da bei bestehenden industriellen Systemen das dafür notwendige Wissen oft nicht vorhanden ist, wurde ein Ansatz entwickelt, der mit Hilfe einer Optimierungsmethode (Simulated Annealing) basierend auf beobachteten Daten ein System *automatisiert* in geeignete Teilsysteme unterteilt. Die Idee des Verfahrens ist es, automatisiert verschiedene Teilsystemunterteilungen zu erstellen, die dann mit Hilfe eines Optimierungskriteriums bewertet werden. Dazu wurden zwei Kriterien entwickelt, die es erlauben den Grad an interner Nebenläufigkeit einer gegebenen Systemunterteilung auf Basis der aufgezeichneten Systemevolutionen näherungsweise zu bestimmen. Durch Minimierung der internen Nebenläufigkeit kann das Optimierungsverfahren selbst Systeme im industriellen Maßstab so unterteilen, dass eine Modellidentifikation nach verhältnismäßig kurzer Beobachtungszeit möglich ist, da die Teilsystemsprachen konvergieren.

Für jedes der Teilsysteme wird ein Automat mit dem monolithischen Identifikationsalgorithmus erstellt. Die so identifizierten Automaten bilden dann ein Automatennetz. Gilt für jede der beobachteten Teilsystemsprachen dass sie zur originalen Teilsystemsprache konvergiert, gilt für die Sprache des Gesamtsystems $L_{Ident}^{k+1} \supseteq L_{Orig}^{k+1}$ (siehe Theorem 6): das Netzwerk aus Automaten kann die gesamte fehlerfreie Systemsprache reproduzieren, selbst wenn Teile davon bisher noch nicht beobachtet wurden. Damit kann die Anzahl der Fehlalarme verglichen mit dem monolithischen Modell erheblich reduziert werden.

Mit der Erzeugung der bisher noch nicht beobachteten fehlerfreien Sprache geht allerdings auch einher, dass das Automatennetzwerk vermehrt Ausgangsfolgen erzeugen kann, die im System nur durch Fehler verursacht werden. Manche Fehler führen beispielsweise dazu, dass jeder der Teilautomaten eine passende Trajektorie findet, um die durch den Fehler verursachte Ausgangsfolge seines Teilsystems zu reproduzieren. Damit kann der Fehler von keinem der Teilautomaten detektiert werden. Der Fehler ist dadurch gekennzeichnet, dass das *kombinierte* Teilsystemverhalten nicht zulässig ist. Daher muss das kombinierte Verhalten des Automatennetzwerks überwacht werden. Bild 9.4 zeigt das Prinzip der Überwachung: über den Teilautomaten, die die Ausgangsfolgen ihrer Teilsysteme reproduzieren, steht eine Struktur bestehend aus einem in dieser Arbeit neu eigeführten Permissive Observed Cross Product (POCP) und einer ebenfalls neu definierten Toleranzspezifikation. Das POCP enthält Informationen darüber, welche kombinierten Trajektorien von Teilautomaten während der Lernphase bereits beobachtet wurden und damit als fehlerfrei gelten können. Darüberhinaus enthält das POCP zusätzliche Transitionen und einen Joker-Zustand, mit dem ermittelt werden kann, ob eine gegebene kombinierte Teilautomatentrajektorie bisher noch nicht beobachtet wurde und damit potentiell auf Grund eines Fehlers auftrat. Mit Hilfe der Toleranzspezifikation ist es möglich den Anteil an akzeptablem neuem kombiniertem Verhalten der Teilautomaten zu definieren. Die Toleranzspezifikation ist ein Automat, der vom Nutzer vorgegeben werden muss. Damit ist es beispielsweise möglich festzulegen, dass eine bisher unbekannte kombinierte Teilsystemtrajektore der Länge zwei toleriert wird, nach Auftreten einer dritten unbekannten Zustandskombination aber ein Fehler detektiert werden soll. Dazu beobachten die Teilautomaten synchronisiert mit dem POCP und der Toleranzspezifikation das System. Die Toleranzspezikation erlaubt es, die Anzahl an Fehlalarmen und die Fehlererkennungsschärfe des Modells auszubalancieren. Fehler können damit auf zwei Ebenen erkannt werden: auf Ebene der Teilsysteme ist es möglich, dass ein Automat das Teilsystemverhalten nicht reproduzieren kann und auf Ebene der Toleranzspezifikation ist es möglich, dass ein kombiniertes Teilsystemverhalten zur Fehlererkennung führt.

Nachdem ein Fehler detektiert wurde, ist der nächste Schritt das Bestimmen der Fehlerursache. Bei Systemen im industriellen Maßstab ist dazu eine Lokalisierung des fehlerhaften Verhaltens unabdingbar. Solche Systeme bestehen oft aus mehreren hundert Sensoren und Aktuatoren. Für eine schnelle Reparatur ist es daher wichtig eine

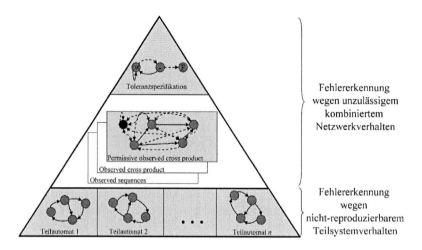

Fehlererkennung
wegen unzulässigem
kombiniertem
Netzwerkverhalten

Fehlererkennung
wegen
nicht-reproduzierbarem
Teilsystemverhalten

Abbildung 9.4: Restriktion des Automatennetzwerks

begrenzte Menge an möglichen Fehlerquellen zu bestimmen. Dies kann beispielsweise in Form von Sensoren oder Aktuatoren erfolgen, die bei der Fehlererkennung ein von der Norm abweichendes Verhalten gezeigt haben. In der betrachteten Klasse von Systemen, bestehend aus Steuerung und Steuerstrecke im geschlossenen Kreis, kann diese Fehlerlokalisierung dadurch erfolgen, dass die Ein- und Ausgänge (E/As) der Steuerung bestimmt werden, die ein nicht reproduzierbares Verhalten gezeigt haben. Da Steuerungs-E/As direkt mit Sensoren und Aktuatoren verbunden sind, kann durch Angabe einer kleinen Menge an möglicherweise fehlerhaften E/As ein defektes Bauteil meist schnell lokalisiert werden.

Bei der Diagnose kontinuierlicher Systeme ist der Einsatz sogenannter Residuen ein weit verbreitetes Mittel, um diejenigen Signale zu ermitteln, die vom Normalverhalten abweichen (Isermann, 2006). Residuen sind Rechenvorschriften, die den Unterschied zwischen modelliertem (also erwartetem) und tatsächlich beobachtetem Verhalten quantifizieren. Ausgehend von dieser Idee wurde in dieser Arbeit eine Methode entwickelt, die es erlaubt den Unterschied zwischen beobachtetem und erwartetem E/A-Verhalten basierend auf Mengenoperationen zu bestimmen. Sowohl für den Diagnoseeinsatz des monolitischen Automaten, als auch für die verteilte Struktur bestehend aus Automatennetzwerk, POCP und Toleranzspezifikation wurden geeignete Operationen definiert, die sowohl das *unerwartete* als auch das *verpasste* E/A-Verhalten bestimmen. Mit Hilfe der so definierten Residuen kann eine relativ kleine Menge an Fehlerkandidaten in Form von Steuerungs-E/As ermittelt werden.

Um beurteilen zu können, ob ein gegebenes Modell für die Online-Diagnose geeignet ist, wurden zwei probabilistische Kenngrößen entwickelt. Sie erlauben es einzuschätzen, wie wahrscheinlich ein fehlerhaftes Verhalten vom Modell reproduziert und damit nicht erkannt werden kann.

Um zu zeigen, dass die in der Arbeit entwickelten Methoden für industrielle Systeme

161

eingesetzt werden können, wurde im Rahmen einer Industriekooperation ein Diagnosesystem für einen industriellen Wickler implementiert. Dabei konnte gezeigt werden, dass insbesondere durch den Einsatz der *verteilten* Identifikation auf Basis automatisch generierter Teilsysteme die für die Diagnose notwendigen Modelle auch für große Systeme mit verhältnismäßig wenig Lerndaten erzeugt werden können. Es konnte gezeigt werden, dass durch den Einsatz der verteilten Modelle die Anzahl der Fehlalarme verglichen mit dem Einsatz des monolithischen Modells signifikant gesenkt wurde, ohne dass die Fähigkeit zur Fehlererkennung übermäßig beeinträchtigt ist.

9.2 Résumé en langue française

La compétitivité des entreprises manufacturières dépend fortement de la productivité des machines et des moyens de production mis en œuvre. C'est pourquoi il est indispensable de minimiser les temps d'arrêt dus aux fautes ou dysfonctionnements. Afin d'atteindre un temps de réparation minimal, il est nécessaire d'acquérir rapidement des informations sur la cause d'une faute apparue dans un système. Le développement de méthodes de diagnostic dans le domaine de l'automatique est motivé par de tels besoins.

De nombreux systèmes industriels peuvent être considérés comme des systèmes à événements discrets (SED). Ils peuvent souvent être représentés par une boucle fermée entre le contrôleur et le processus (figure 9.5). Le comportement de tels systèmes peut être analysé en considérant les signaux échangés entre contrôleur et processus. Le but de ce travail est de développer une approche de diagnostic générique pour cette classe de systèmes.

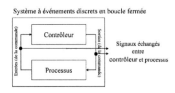

FIGURE 9.5: Systèmes à événements discrets en boucle fermée

La plupart des approches de diagnostic pour les systèmes techniques peuvent être classifiées dans l'une des trois catégories suivantes : *approches basées sur des données*, *système expert* ou *approches basées sur des modèles*. Une analyse de ces trois catégories montre que les approches basées sur des modèles ont le potentiel le plus grand pour être appliquées aux SED en boucle fermée : L'idée de ces approches consiste en la comparaison du comportement observé avec le comportement attendu par un modèle de référence (figure 9.6). En comparant les deux comportements, il est possible de décider si la situation actuelle du système est symptomatique d'une faute ou si elle représente le bon fonctionnement. Les approches basées sur des modèles sont particulièrement bien adaptées à la classe de systèmes considérée car un modèle permet de représenter la dynamique d'un système. Avec les modèles pour les SED (automates, réseaux de Petri) il est possible de modéliser l'état d'un système. Selon cet état, un comportement donné peut être soit le résultat du bon fonctionnement soit le résultat d'une faute. Utiliser des modèles à état permet souvent de classifier plus précisément un comportement observé qu'en travaillant avec des approches basées sur des données ou avec des systèmes expert.

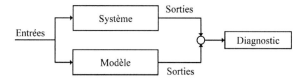

FIGURE 9.6: Principe du diagnostic basé sur des modèles

Dans la littérature scientifique, on trouve deux types d'approches basées sur des modèles. Le premier type consiste en des méthodes utilisant un modèle qui contient à la fois le comportement normal et le comportement fautif. Si un comportement observé est conforme à un comportement fautif modélisé, la faute correspondante est détectée. L'inconvénient de ce type de méthodes est que seules les fautes menant à un comportement explicitement inclus dans le modèle peuvent être détectées et localisées. Le deuxième type de méthodes permet d'éviter cet inconvénient. Il s'agit des approches basées sur des modèles ne contenant que le comportement normal. Une faute est détectée dès qu'un comportement observé ne peut pas être reproduit par le modèle. Cependant, dans ce cas, les modèles contiennent moins de savoir sur le système et trouver la cause d'une faute est souvent plus difficile qu'avec des méthodes basées sur des modèles incluant les comportements fautifs.

En travaillant avec des méthodes basées sur des modèles, un défi important est la construction des modèles. Modéliser manuellement des systèmes de taille industrielle est coûteux. C'est pourquoi ce travail vise une approche d'identification. Dans un premier temps, l'identification des automates monolithiques a été considérée. L'identification est basée sur des données du système observé pendant des évolutions sans fautes. Ce type d'automate non-déterministe autonome avec fonction de sortie fonctionne comme un générateur d'événements comparable à la classe de SED en boucle fermée. La première contribution de ce travail est d'améliorer un algorithme d'identification proposé par (Klein, et al. 2005). Cet algorithme comporte un paramètre $k \in \mathbb{N}^+$ pour ajuster l'exactitude du modèle par rapport aux comportements observés. L'algorithme fournit un automate qui est $k+1$-complet. C'est-à-dire son langage[2] L_{Ident}^{k+1} est exactement égal au langage observé de longueur $k+1$: $L_{Ident}^{k+1} = L_{Obs}^{k+1}$. Si une faute mène à une séquence $w \notin L_{Obs}^{k+1}$, alors $w \notin L_{Ident}^{k+1}$ pendant le diagnostic en ligne avec le modèle identifié, il est ainsi garanti que l'automate ne peut pas reproduire cette séquence. Par conséquent il est assuré que la faute est détectée.

Une condition importante pour bien choisir la valeur du paramètre k est la relation entre le langage observé L_{Obs}^{k+1} et le langage du bon fonctionnement du système L_{Orig}^{k+1}. Après que les données du système aient été collectées, il est important d'avoir

---

[2]Un langage L^n est composée de séquences de symboles de sortie de longueur inférieure ou égale à n (ici : de séquences de vecteurs d'entrées et sorties du contrôleur)

une évolution de L_{Obs}^{k+1} qui permet de constater que $L_{Obs}^{k+1} = L_{Orig}^{k+1}$. C'est-à-dire il est important d'avoir observé tous les comportements normaux possibles du système. Si $L_{Obs}^{k+1} \subset L_{Orig}^{k+1}$, il y aura toujours une fausse alerte pendant le diagnostic en ligne quand une séquence $w \in L_{Orig}^{k+1}$ apparaîtra et ne pourra pas être reproduite par l'automate car $w \notin L_{Obs}^{k+1} = L_{Ident}^{k+1}$. La convergence de L_{Obs}^{k+1} vers L_{Orig}^{k+1} comme décrit dans la figure 9.7 peut servir comme critère pour constater que $L_{Obs}^{k+1} = L_{Orig}^{k+1}$. Si le système est observé suffisamment longtemps, il montre de plus en plus de son langage orignal qui peut ensuite être inclus dans le modèle. Si tout le comportement original (c'est-à-dire le comportement du bon fonctionnement) est apparu pendant l'observation des données pour l'identification, on va constater que L_{Obs}^{k+1} ne croît plus même lorsque le temps d'observation continue d'augmenter. C'est-à-dire que si $|L_{Obs}^{k+1}|$ (le cardinal de l'ensemble) converge, il peut raisonnablement être envisagé que $L_{Obs}^{k+1} = L_{Orig}^{k+1}$.

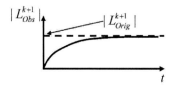

FIGURE 9.7: Convergence du langage observé

Il a cependant été constaté pendant ce travail que cette convergence peut prendre très longtemps pour des systèmes comportant un parallélisme prononcé. La raison est que de tels systèmes sont capables de produire un très grand nombre de comportements qui doivent être collectés pendant la phase d'observation du système. Une observation incomplète des comportements du système mène à un grand nombre de fausses alertes qui rend la méthode de diagnostic inutilisable. En raison de contraintes économiques, souvent il n'est pas possible d'augmenter le temps d'observation jusqu'à ce que le langage observé converge. Pour résoudre ce problème, une approche d'identification distribuée a été développée dans ce travail. L'idée de base de l'approche est de diviser un système en sous-systèmes avec peu de parallélisme interne. Si les sous-systèmes ont été soigneusement choisis, leurs langages observés convergent plus vite que le langage du système global. Une possibilité pour trouver une partition appropriée est d'utiliser le savoir des experts du système. Puisque dans beaucoup de systèmes industriels ce savoir n'est pas disponible pour un coût raisonnable, une approche a été développée pour obtenir automatiquement les sous-systèmes à partir du langage observé. L'approche utilise la technique d'optimisation du recuit simulé. L'idée de cette approche est de créer différentes partitions possibles du système et de les évaluer à l'aide de critères d'optimisation. Deux critères d'optimisation ont été développés et permettent de déterminer approximativement le degré de parallélisme interne d'une partition donnée. En minimisant le parallélisme interne, l'approche d'optimisation peut même partitionner des systèmes de taille industrielle afin que les langages des sous-systèmes convergent. Par conséquent cette méthode permet d'identifier des modèles pour des systèmes industriels après un temps d'observation relativement court.

Pour chaque sous-système un automate peut être identifié avec l'algorithme d'identification monolithique. Les automates forment un réseau. Si les langages observés des sous-systèmes convergent tous vers les langages originaux, le langage créé par le réseau d'automates est un simulateur pour le langage original du système entier : $L_{Ident}^{k+1} \supseteq L_{Orig}^{k+1}$. C'est-à-dire : le réseau d'automates peut reproduire tout le comportement du bon fonctionnement du système entier même si ce comportement a été observé seulement de manière partielle. Par conséquent il est possible de réduire le nombre de fausses alertes de manière significative par rapport aux résultats obtenus avec le modèle monolithique.

La capacité à produire le langage original non encore observé s'accompagne du phénomène que le réseau d'automates produit plus de séquences de sorties qui représentent des fautes dans le système. Il existe des fautes qui mènent à une situation dans lequel chaque automate partiel trouve une trajectoire pour reproduire la séquence de sortie de son sous-système. Aucun des automates partiels n'est capable de détecter une telle faute. La faute est caractérisée par un comportement combiné des sous-systèmes qui n'est pas acceptable. C'est pourquoi le comportement combiné du réseau d'automates doit être surveillé. La figure 9.8 montre le principe de cette surveillance : au dessus du réseau d'automates se trouve une structure composée d'un Permissive Observed Cross Product (POCP) et d'une spécification de tolérance qui ont été introduits dans ce travail. Le POCP contient des informations sur les trajectoires du réseau d'automates observées pendant la phase d'apprentissage. Ces trajectoires peuvent être considérées non fautives puisqu'elles ont été observées pendant des évolutions du système sans fautes. Avec le POCP il est possible de décider si une trajectoire du réseau d'automates est connue ou bien si elle n'a pas encore été observée et est donc susceptible de représenter une faute. A l'aide de la spécification de tolérance il est possible de limiter la part du comportement inconnu qui est acceptée comme normale et qui ne mène pas à la détection d'une faute. La spécification de tolérance est un automate qui doit être défini a priori. Avec cette spécification il est par exemple possible de définir qu'une trajectoire inconnue de longueur deux peut être tolérée mais qu'à partir d'une longueur trois une faute doit être détectée. La spécification de tolérance permet d'équilibrer le nombre de fausses alertes et la précision du modèle par rapport aux comportements fautifs. Avec la structure décrite, il est possible de détecter des fautes à deux niveaux : au niveau du réseau d'automates il est possible qu'un automate ne soit pas capable de reproduire le comportement de son sous-système et au niveau de la spécification de tolérance il est possible qu'un comportement combiné des sous-systèmes dépasse la limite définie.

Après qu'une faute ait été détectée il est nécessaire de déterminer sa cause. Dans ce contexte, les systèmes de taille industrielle exigent une localisation de la faute. Ces systèmes comportent souvent quelques centaines d'actionneurs et de capteurs. C'est pourquoi il est indispensable de déterminer un ensemble limité de candidats de fautes, c'est-à-dire un ensemble de composants du système qui ont montré un comportement anormal. Dans la classe des SED en boucle fermée du contrôleur et du processus, la localisation de ces composants peut être réalisée en donnant des entrées ou des sorties (E/S) du contrôleur qui ont montré un comportement non reproductible. Puisque les E/S du contrôleur sont directement connectées avec les actionneurs ou les capteurs, il est souvent possible de trouver un composant défectueux en analysant un petit ensemble

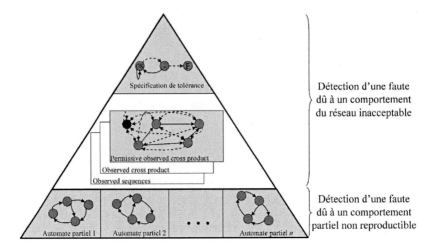

FIGURE 9.8: Surveillance du réseau d'automates

d'E/S potentiellement reliées à la faute détectée.

Un concept bien connu pour le diagnostic des systèmes continus est d'utiliser des résidus pour déterminer des signaux qui divergent du comportement normal (Isermann, 2006). Les résidus consistent en des procédures de calcul pour quantifier la différence entre le comportement observé et le comportement attendu (modélisé). Basée sur ce concept, une nouvelle méthode pour déterminer la différence entre le comportement attendu et le comportement observé d'E/S du contrôleur a été conçue dans ce travail. Pour le modèle monolithique et pour le modèle composé du réseau d'automate, du POCP et de la spécification de tolérance, des procédures de calculs ont été développés pour déterminer les E/S qui montrent un comportement inattendu ou qui ne montrent pas un comportement attendu par le modèle. A l'aide de ces procédures il est possible de déterminer un ensemble limité d'E/S qui sont potentiellement reliées à un composant fautif.

Afin de juger si un modèle donné est bien adapté au diagnostic en ligne, deux mesures probabilistes ont été définies. Elles permettent d'estimer la probabilité qu'un comportement fautif peut être reproduit par le modèle et par conséquent n'est pas détectable.

Une application industrielle de la méthode développée dans ce travail a été traitée afin de montrer que l'approche est applicable à l'échelle industrielle. Dans le cadre d'une coopération avec un fabriquant de textile non-tissé, un système de diagnostic pour un enrouleur industriel a été implémenté. Il a été possible de montrer que l'identification distribuée basée sur des sous-systèmes obtenue avec l'approche d'optimisation permet de créer de bons modèles avec relativement peu de données. Il a également été montré que l'application du réseau d'automates avec POCP et spécification de tolérance réduit le nombre de fausses alertes de manière significative sans pour autant que la capacité à détecter des fautes soit trop fortement dégradée.

10 Nomenclature

10.1 Abbreviations

Abbreviation	Explanation	Page
DES	Discrete Event System	
NDAAO	Non-Deterministic Autonomous Automaton with Output	41
POCP	Permissive Observed Cross Product	

10.2 Variables

Variable	Explanation	Page
L_{Orig}	Original (fault-free) language of a closed-loop DES	28
L_{Obs}	Observed language of a closed-loop DES	28
L_{Obs}^n	Observed language of length n of a closed-loop DES	42
L_{Obs,sys_t}^n	Observed partial language of subsystem sys_t	66
$L_{Obs,sys_t}^{n,h}$	Observed partial language of subsystem sys_t based on the first h of p system evolutions	98
W_{Obs}^n	Observed words of length n of a closed-loop DES	42
L_{Ident}	Identified language of a closed-loop DES (language of the DES model)	28
L_{Ident}^n	Identified language of length n of a closed-loop DES	43
W_{Ident}^n	Identified words of length n of a closed-loop DES	43
L_{Exc}	Exceeding language of a closed-loop DES ($L_{Exc} = L_{Ident} \backslash L_{Orig}$)	28
L_{NR}	Non-reproducible language of a closed-loop DES ($L_{NR} = L_{Orig} \backslash L_{Ident}$)	28
u^{DES}	Output of a closed-loop DES	40
X^{DES}	Closed-loop DES state space	40
x^{DES}	State of a closed-loop DES	40
x_0^{DES}	Initial state of a closed-loop DES	40
X	State space of an NDAAO	41
x_0	Initial state of an NDAAO	41
Ω	Output alphabet of an NDAAO	41
$u_h(j)$	j-th output of in the h-th of p system evolutions	41

Variable	Explanation	Page
σ_h	Observed sequence during the h-th system evolution	42
Σ	Set of all observed sequences	42
u	Controller I/O vector (precise definition of $u_h(j)$ for closed-loop DES)	66
IO_i	Controller I/O	66
sys_t	t-th subsystem	66
$y(sys_t)$	I/O mapping function (assigns a set of controller I/Os to subsystem sys_t)	66
N_{sys}	Number of subsystems	66
u_{sys_t}	Partial controller I/O vector (I/O which are not part of sys_t are replaced by '-')	66
$NDAAO_{\|\|}$	NDAAO cross product	69
$X_{\|\|}$	State space of an NDAAO cross product	69
$x_{\|\|0}$	Initial state of an NDAAO cross product	69
$NDAAO_{Tol}$	Tolerance specification automaton	75
X_{Tol}	State space of a tolerance specification	75
x_{Tol_0}	Initial state of a tolerance specification	75
Ω_{Tol}	Output alphabet of a tolerance specification	75
$NDAAO_{Obs\|\|}$	Observed cross product	78
$X_{Obs\|\|}$	State space of the observed cross product	78
$x_{Obs\|\|0}$	Initial state of the observed cross product	78
$\Omega_{Obs\|\|}$	Output alphabet of the observed cross product (state combinations of the underlying partial automata)	78
X_{POCP}	State space of the permissive observed cross product	80
x_{POCP}	Initial state of the permissive observed cross product	80
Ω_{POCP}	Output alphabet of the permissive observed cross product (state combinations of the underlying partial automata)	80
EC_1, EC_2	Evaluation criteria	109
E	Set of controller I/O edges	119
$\widetilde{X}_t, \widetilde{X}_{t-1}$	Current and former states estimation	121
T_{Obs}	Observed trajectory after the tolerance specification left its OK state	134

10.3 Functions

Function	Explanation	Page
F_N	Non-deterministic next state function of closed-loop DES	40
Λ	Output function of a closed-loop DES ($u^{DES} = \Lambda(x^{DES})$)	40
f	Non-deterministic transition function of an NDAAO	41
λ	Output function of an NDAAO	41
$\widetilde{\lambda}$	Output function of an identified NDAAO until step 4 of algorithm 1, delivers a word	48

Function	Explanation	Page
Gap	Transition state gap	57
J	Join function to combine two partial I/O vectors (delivers 'c' for I/Os which are contradictory)	68
$f_{\parallel}$	Non-deterministic transition function of an NDAAO cross product	69
$\lambda_{\parallel}$	Output function of an NDAAO cross product	69
RemEqual	Replace equal substrings function (with $w^5 = ABBBC$, RemEqual(w^5)=ABC)	70
IOProj$_{IOSet}$	I/O vector projection to the I/O list $IOSet$ (is applied to an I/O vector: Each I/O not belonging to $IOSet$ is replaced by '-')	70
Θ	Transition observation function (true if a transition has been observed during identification)	73
f_{Tol}	Transition function of a tolerance specification	75
λ_{Tol}	Output function of a tolerance specification	75
Θ_{Tol}	Transition observation function for the tolerance specification	75
$f_{Obs\parallel}$	Transition function of the observed cross product	78
$\lambda_{Obs\parallel}$	Output function of the observed cross product (delivers the combination of underlying partial automata states of a given $x_{Obs\parallel}$)	78
f_{POCP}	Transition function of the permissive observed cross product	80
λ_{POCP}	Output function of the permissive observed cross product (delivers the combination of underlying partial automata states of a given x_{POCP} or ε if it is the 'joker'-state)	80
Θ_{POCP}	Transition observation function for the POCP	80
$CausalMap$	Causal actuator sensor map	106
$Edge$	Edge function (to get the rising or falling edge when comparing two controller I/O values)	119
ES	Evolution set (to get the rising or falling edges between two I/O vectors)	120
FD	Fault detection policy to decide if a fault is detected	121
ES_T	Evolution set for the distributed model (to get the rising or falling edges in an I/O vector trajectory)	136
J_{POCP}	Join function for the POCP	136
$\widetilde{\Psi}$	Determines the expected trajectories in the POCP	136

10.4 Operators

Operator	Explanation	Page
() as $x(j)$	State after the j-th event	
() as $u(j)$	Output after the j-th event	
$\|Set\|$	Set cardinality (e.g. $\|f(x)\|$ for the number of following states of x)	
$\|Vector\|$	Length of the vector (number of elements)	
2^{Set}	Powerset of Set	40
$\langle .. \rangle$	Substring selection operator (with $w^q = ABCDEF$ follows $w^q\langle 2..4 \rangle = BCD$)	44
[ ]	Selection of elements in a vector: $u[i]$ selects the i-th element of vector u	68

Bibliography

Alur, R. (1999). Timed automata. In *Computer Aided Verification*, volume 1633 of *Lecture Notes in Computer Science*. Springer Berlin / Heidelberg.

ANSI/IEEE100 (1997). *The IEEE Standard Dictionary of Electrical and Electronics Terms according to ANSI/IEEE standard 100-1988*. IEEE Standards Office, New York.

Baron, C., Geffroy, J., and Zamilpa, C. (2001a). Identification of evolutionary sequential systems - part 1: unified approach. *Communications in numerical mehtods in engineering*, 17(9):623–630.

Baron, C., Geffroy, J., and Zamilpa, C. (2001b). Identification of evolutionary sequential systems - part 2: genetic simulation experiments. *Communications in numerical mehtods in engineering*, 17(9):631–637.

Biermann, A. and Feldman, J. (1972). Synthesis of finite-state machines from samples of their behavior. *IEEE Transactions on Computers*, 21(6):592–597.

Blanke, M., Kinnaert, M., Lunze, J., and Staroswiecki, M. (2006). *Diagnosis and Fault-Tolerant Control*. Springer Berlin Heidelberg, 2nd edition.

Booth, T. (1967). *Sequential Machines and Automata Theory*. John Wiley & Sons Inc, Hoboken (New Jersey).

Bérard, B., Bidoit, M., Finkel, A., Laroussinie, F., Petit, A., Petrucci, L., and Schnoe-belen, P. (2001). *Systems and Software Verification*. Springer Berlin / Heidelberg.

Cassandras, C. G. and Lafortune, S. (2006). *Introduction to Discrete Event Systems*. Springer New York.

Chen, J. and Patton, R. (1998). *Robust model-based fault diagnosis for dynamic systems*. Springer Berlin / Heidelberg.

Cook, J., Du, Z., Liu, C., and Wolf, A. (2004). Discovering models of behavior for concurrent workflows. *Computers in Industry*, 53(3):297–319.

Cordier, M.-O. and Dousson, C. (2000). Alarm driven monitoring based on chronicles. In *Proceedings of the 4th Symposium on Fault Detection Supervision and Safety for Technical Processes (Safeprocess 2000)*, pages 286–291.

Dash, D. and Venkatasubramanian, V. (2000). Challenges in the industrial applications of fault diagnostic systems. *Computers & Chemical Engineering*, 24(2-7):785–791.

Debouk, R., Lafortune, S., and Teneketzis, D. (2000). Coordinated decentralized protocols for failure diagnosis of discrete event systems. *Discrete Event Dynamic Systems*, 10(1-2):33–86.

Dotoli, M., Fanti, M. P., and Mangini, A. M. (2006). On-line identification of discrete event systems: a case study. In *Proccedings of the 2006 IEEE International Conference on Automation Science and Engineering*, pages 405–410.

Dotoli, M., Fanti, M. P., and Mangini, A. M. (2008). Real time identification of discrete event systems using Petri nets. *Automatica*, 44(5):1209–1219.

Dotoli, M., Fanti, M. P., Mangini, A. M., and Ukovich, W. (2009). On-line fault detection in discrete event systems by petri nets and integer linear programming. *Automatica*, 45(11):2665–2672.

Franklin, G. F., Powell, D. J., and Emami-Naeini, A. (2001). *Feedback Control of Dynamic Systems*. Prentice Hall.

FreudenbergGroup (2009). Annual report 2009. Technical report, Freudenberg Group.

Frey, G. and Younis, M.-B. (2004). A re-engineering approach for PLC programs using finite automata and UML. In *Proceedings of the 2004 IEEE International Conference on Information Reuse and Integration (IRI2004)*, pages 24–29.

Genc, S. and Lafortune, S. (2003). Distributed diagnosis of discrete-event systems using petri nets. *Lecture Notes in Computer Science*, 2679:316–336.

Genc, S. and Lafortune, S. (2007). Distributed diagnosis of place-bordered Petri nets. *IEEE Transactions on Automation Science and Engineering*, 4(2):206–219.

Ghallab, M. (1996). On chronicles: Representation, on-line recognition and learning. In *Proceedings of the 5th International Conference on Principles of Knowledge Representation and Reasoning (KR '96)*, pages 597–606.

Giua, A. and Seatzu, C. (2005). Identification of free-labeled petri nets via integer programming. In *Proceedings of the 44th IEEE Conference on Decision and Control, and the European Control Conference*, pages 7639–7644.

Hashtrudi Zad, S., Kwong, R. H., and Wonham, W. M. (2005). Fault diagnosis in discrete-event systems: Incorporating timing information. *IEEE Transactions on Automatic Control*, 50(7):1010–1015.

IEC (2002). Grafcet - specification language for sequential function charts IEC 60848.

Isermann, R. (2006). *Fault-Diagnosis Systems: An Introduction from Fault Detection to Fault Tolerance*. Springer Berlin-Heidelberg.

Isermann, R. and Balle, P. (1997). Trends in the application of model-based fault detection and diagnosis of technical processes. *Control engineering practice*, 5(5):709–719.

Kella, J. (1971). Sequential Machine Identification. *IEEE Transactions on Computers*, 20(3):332–336.

Klein, S. (2005). *Identification of Discrete Event Systems for Fault Detection Purposes*. Shaker Verlag, Aachen.

Korbicz, J., Koscielny, J.-M., Kowalczuk, Z., and W., C., editors (2004). *Fault Diagnosis: Models, Artificial Intelligence, Applications*. Springer Berlin/Heidelberg, 1st edition.

Lee, A. E. and Varaiya, P. (2002). *Structure and Interpretation of Signals and Systems*. Addison Wesley, United States.

Machado, J., Denis, B., and Lesage, J. J. (2006). A generic approach to build plant models for DES verification purposes. In *Proceedings of the 8th International Workshop on Discrete Event Systems (WODES 2006)*, pages 407–412.

Maruster, L., Weijters, A., van den Bosch, A., and Daelemans, W. (2003). Discovering process models by rule set induction. In *Proceedings of the 5th International Workshop on Symbolic and Numeric Algorithms for Scientific Computing*, pages 180–191.

Meda-Campana, M. E. and Lopez-Mellado, E. (2005). Identification of concurrent discrete event systems using petri nets. In *Proceedings of the Mathematical Computer, Modelling and Simulation Conference (IMACS 2005)*.

Michalewisz, Z. and Fogel, D. B. (2000). *How to Solve it: Modern Heuristics*. Springer Verlag.

Moor, T., Raisch, J., and Young, S. (1998). Supervisory control of hybrid systems via l-complete approximations. In *Proceedings of the IEEE fourth Workshop on Discrete Event Systems WODES 98*, pages 426–431.

Murata, T. (1989). Petri nets: Properties, analysis and applications. *Proceedings of the IEEE*, 77(4):541–580.

Neidig, J. and Lunze, J. (2005). Decentralised diagnosis of automata networks. In *Proceedings of the 16th IFAC World Congress*, Prague.

Neidig, J. and Lunze, J. (2006). Unidirectional coordinated diagnosis of automata networks. In *Proceedings of the 17th International Symposium on Mathematical Theory of Networks and Systems*, pages 2203–2208.

Pandalai, D. and Holloway, L. (2000). Template languages for fault monitoring of timed discrete event processes. *IEEE transactions on automatic control*, 45(5):868–882.

Papadopoulos, Y. and McDermid, J. (2001). Automated safety monitoring: A review and classification of methods. *International Journal of Condition Monitoring and Diagnostic Engineering Management*, 4(4):14–32.

Papoulis, A. and Unnikrishna, P. (2002). *Probabilistic, Random Variables and Stochastic Processes*. McGraw-Hill.

Philippot, A. (2006). *Contribution au diagnistic décentralisé des systèmes à événements discrets : application aux systèmes manufacturiers*. PhD thesis, Université de Reims Champagne-Ardenne.

Philippot, A., Sayed-Mouchaweh, M., and Carré-Ménétrier, V. (2005). Diagnostic des SED par modélisation multi-outils. In *Proceedings of the Journées Doctorales MACS*.

Philippot, A., Sayed-Mouchaweh, M., and Carré-Ménétrier, V. (2007). Unconditional decentralized structure for the fault diagnosis of discrete event systems. *Proceedings of the 1st IFAC Workshop on Dependable Control of Discrete Systems DCDS'07*.

Pia Fanti, M. and Seatzu, C. (2008). Fault diagnosis and identification of discrete event systems using petri nets. In *Proceedings of the 9th International Workshop on Discrete Event Systems (WODES)*, pages 432–435.

Reiter, R. (1987). A theory of diagnosis from first principles. *Artificial Intelligence*, 32(1):57–95.

Riordan, J. (2002). *An Introduction to Combinatorial Analysis*. Dover Publications, New York.

Roth, M., Jean-Jacques, L., and Litz, L. (2009a). Distributed identification of discrete event systems for fault detection purposes. In *Proceedings of the European Control Conference 2009*, pages 2590–2595.

Roth, M., Jean-Jacques, L., and Litz, L. (2009b). An FDI method for manufacturing systems based on an identified model. In *Proceedings of the 13th IFAC Symposium on Information Control Problems in Manufacturing, INCOM'09*, pages 1389–1394.

Roth, M., Lesage, J.-J., and Litz, L. (2009c). A residual inspired approach for fault localization in DES. In *Proceedings of the 2nd IFAC Workshop on Dependable Control of Discrete Event Systems (DCDS'09)*, pages 347–352.

Roth, M., Lesage, J.-J., and Litz, L. (2010). Identification of discrete event systems - implementation issues and model completeness. In *Proceedings of the 7th International Conference on Informatics in Control, Automation and Robotics (ICINCO)*, pages 73–80.

Sampath, M., Sengupta, R., Lafortune, S., Sinnamohideen, K., and Teneketzis, D. (1996). Failure diagnosis using discrete-event models. *IEEE transactions on control systems technology*, 4(2):105–124.

Sayed-Mouchaweh, M., Philippot, A., and Carre-Menetrier, V. (2008). Decentralized diagnosis based on Boolean discrete event models: application on manufacturing systems. *International journal of production research*, 46(19):5469–5490.

Schroder, J. (2002). *Modelling, State Observation, and Diagnosis of Quantised Systems.* Springer-Verlag New York.

Supavatanakul, P., Lunze, J., Puig, V., and Quevedo, J. (2006). Diagnosis of timed automata: Theory and application to the damadics actuator benchmark problem. *Control Engineering Practice*, 14(6):609–619.

Tornambe, A. (1996). *Discrete-Event System Theory: An Introduction.* World Scientific Publishing Company.

van Schuppen, J. H. (2004). System theory for system identification. *Journal of Econometrics*, 118:331–339.

Veelenturf, L. (1978). Inference of Sequential-Machines from Sample Computations. *IEEE Transactions on Computers*, 27(2):167–170.

Venkatasubramanian, V., Rengaswamy, R., Kavuri, S., and Yin, K. (2003). A review of process fault detection and diagnosis Part III: Process history based methods. *Computers & Chemical Engineering*, 27(3):327–346.